KU-791-720

# A GUIDE TO INFORMATION SOURCES IN THE GEOGRAPHICAL SCIENCES

# A Guide to Information Sources in the Geographical Sciences

**Edited by
Stephen Goddard**



CROOM HELM
London & Canberra

BARNES & NOBLE BOOKS
Totowa, New Jersey

Croom Helm Ltd, Provident House, Burrell Row,
Beckenham, Kent BR3 1AT
Croom Helm Australia Pty Ltd, 28 Kembla,
Fyshwick, ACT 2609, Australia

British Library Cataloguing in Publication Data

Goddard, Stephen
A guide to information sources in the geographical sciences.
1. Geography – Information services
I. Title
910'.7 G64

ISBN 0-7099-1150-5

First published in the USA 1983 by
Barnes and Noble Books,
81 Adams Drive,
Totowa, New Jersey, 07512

Library of Congress Cataloging in Publication Data
Main entry under title:

A Guide to information sources in the geographical sciences.

1. Geography – Bibliography. 2. Geography – Information services. I. Goddard, Stephen.
Z6001.G84 1983 [G116] 016.91 83-6065
ISBN 0-389-20403-X

Printed and bound in Great Britain

CONTENTS

Tools for the Geographer

## FIGURES AND TABLES

### Figures



### Tables

## CONTRIBUTORS

J.A. ALLAN, B.A., Ph.D. Senior Lecturer, Department of Geography, School of Oriental and African Studies, University of London.

W.J. ALSPAUGH, B.A., Assistant Bibliographer and Supervisor, South Asia Collection, Joseph Regenstein Library, University of Chicago.

W.J. CAMPBELL, B.Sc., Lecturer, Department of Geography, University College, University of London.

STEPHEN GODDARD, M.A., A.L.A., Deputy Librarian, Ealing College of Higher Education, London.

R.E. GRIM, B.A., M.A., Ph.D., Bibliographer, Library of Congress, Washington (D.C.).

C.D. HARRIS, M.A., Ph.D., Professor, Department of Geography, University of Chicago.

G.R.P. LAWRENCE, M.Sc., Senior Lecturer, Department of Geography, King's College, University of London.

W.B. MORGAN, M.A., Ph.D., Professor, Department of Geography, King's College, University of London.

HAZEL M. ROBERTS, B.A., Dip.Lib., A.L.A., Librarian and Teacher with experience in Zambia and Tanzania.

The Late G.T. WARWICK, M.B.E., B.Sc., Ph.D., Lately Reader, Department of Geography, University of Birmingham.

J.H. WHEELER, Jr., B.Sc., Ph.D., Professor, Department of Geography, University of Missouri-Columbia.

P.A. WOOD, B.Sc., Ph.D., Lecturer, Department of Geography, University College, University of London.

J.A. YELLING, B.A., Ph.D., Lecturer, Department of Geography, Birkbeck College, University of London.

# PREFACE

Just as any subject can be seen to have an historical component, so any subject has a geographical one - a component of space and place. Everything has its place in a time continuum - there is nothing that can be fully understood without its being seen in this context. Similarly everything that has taken place, is taking place or will take place on the surface of the Earth has a locational significance. This factor is the key of the geographical sciences.

The geographical component of knowledge can be approached in two different ways. In the first place it can be looked at systematically. A given phenomenon can be considered and its spatial significance be looked into at depth. For example we may consider the occurrence of thunderstorms and why their ferocity and frequency differ from place to place. We may consider the areal distribution of poverty - or for that matter of wealth - and consider what factors led to the particular pattern that emerges and what can be done to make things more egalitarian or not - depending on your political beliefs. Secondly the problem can be looked at regionally. We can consider the interrelationship of innumerable factors on a given portion of the Earth's surface. We can consider the world mathematically by looking at the Northern hemisphere or the Southern hemisphere - the Eastern hemisphere or the Western hemisphere. We may consider it on a continental basis or as is more usual, on a political basis. At a macro-scale we may consider individual countries and amalgamations such as the European Economic Community (EEC) or the Economic Community of West African States (ECOWAS). On the micro-scale counties, or individual metropolitan and rural areas might be considered. We may even consider all the special interrelationships in an individual copice or in a particular agricultural hamlet. The possibilities are legion.

Sections one and two of this guide to bibliographic resources attempt to reflect this dual nature of geographical science. Firstly the resources in various typical systematic areas are considered. Secondly certain parts of the Earth's surface have been selected - namely the continent of Africa, the sub-continent of India (or as it is more commonly known today, South Asia) and the political giants - the USSR and the USA. To these a third section has been added that deals with certain tools that are going to be used by persons working in the geographical sciences, systematically or regionally - maps, aerial photographs, statistical material and archives. The book is a guide to the bibliographical resources available to the prospective worker in the geographical component of knowledge; be he or she student, teacher, librarian or whatever.

However in no way can it be looked upon as a comprehensive guide to the literature in the field of the geographical sciences. In the first place there are a vast number of systematic areas of study that are not adequately considered. I have spoken earlier of the geography of electric storms. Climate plays a very important role in a study of man's environment and is not considered here in any specific way. An area of ever increasing importance today is Medical Geography - this too is not looked at in its own right. Again in the section of the book on regional studies it has been necessary to be selective. However the selection is in no way arbitrary. Care has been taken to select developed and developing areas, areas of free enterprise, those with a state controlled economy and also areas that are commonly described as belonging to the Third World. If we use the terminology that has come to be associated with the Brandt Report we have included countries from North, South, East and West. Obviously large parts of the globe have not been covered. The reader wishing to be introduced to the literature of such areas as East Asia (still often referred to unfortunately as the Far East) or Australasia for example will not find their areas covered directly. However it is hoped that they will find the chapters that are included useful in getting to know the kind of literature they need to explore and where they are likely to find it. Many of the works referred to here, such as the Research Catalogue of the American Geographical Society and Harris and Fellman's International List of Geographical Serials carry references to material relating to East Asia and the Antipodes just as they do material on Africa, South Asia, the U.S.A. and the U.S.S.R.

Buckhurst Hill

# ACKNOWLEDGEMENTS

The idea to publish a guide to information sources in the geographical sciences would have remained no more than an idea but for the work of the various teachers and librarians who have contributed to this volume. To all of them I express my sincere thanks. Likewise I should like to thank Croom Helm for accepting the responsibility of publishing the work and particularly Peter Sowden who has displayed remarkable patience!

For permission to reproduce Figure 1, which appeared in G.R. Crone Maps and their Makers (4th edn., 1970), I have to thank Hutchinson's and the author. Tables 2 and 3 are reproduced from G.R.P. Lawrence Cartographic Methods (1979) and I thank Methuen Publications for permission to do so. Figures 7 and 8 and Table 6 are taken from the United States Geological Survey Landsat Data User's Handbook (1978) and I have Paul F. Clarke, Publications Officer of the USGS in Reston, Virginia, to thank for authorising this. Figure 6 and Tables 5 and 7 come from Remote Sensing: Principles and Interpretation (1978) by Floyd F. Sabins Jr. and the copyright is held by W.H. Freeman and Company of San Francisco, California. Mrs Marilyn Kelso, Permissions Officer, kindly gave permission to use them here.

All cartographic work has been undertaken by Miss Liz Johnson and she is sincerely to be thanked. For secretarial work Mrs Irene Cummings must similarly be thanked - from the earliest letters to prospective contributors until the final typing of the text her assistance has been invaluable. Various colleagues have offered advice and assistance for which I am grateful. Professor B.W. Hodder and Chris Perkins of the School of Oriental and African Studies deserve special mention. Miss Julie Wilcocks from the Library of the University of Witwatersrand is also to be thanked for the advice she gave while on study leave in the U.K. recently. Mr Douglas Foskett, Director of the University of London Library has supported this project throughout and to him I must offer special thanks.

Chapter One

# THE COMMUNICATION OF MAN'S SPATIAL EXPERIENCE

Stephen Goddard

## The Geographical Sciences:- an historical note

Science is systematic and formulated knowledge based firmly on the comparison of numerous observations. The geographical sciences may be described as just such a body of knowledge concerning the location and arrangement of differing phenomena on the earth's surface. The collection of observations for the study of man's habitat dates back to earliest times. In the 5th century B.C. Hippocrates wrote his *Of Airs, Waters and Places*, a dissertation on the physical environment of man propounding that it was largely these factors that determined the course of human history. Some 400 years later Ptolemy wrote his *Geographia* which included a treatise on projections for a world map based on a list of the latitude and longitude of some eight thousand locations. These works may be looked upon as the very foundation stones of geography as we know it today. It is on this basis that the subject has grown to produce such sophisticated works as *The Human Experience of Space and Place* by Anne Buttimer and David Seamon with a preface by Professor Hagerstrand (Croom Helm, 1980). And as an aside Professor Hagerstrand might note that despite the prejudices of my profession I know just where to place this valuable little book! The work certainly influenced the author's selection of a title for this chapter.

Following the decline of Greek civilization the collection of observations regarding the surface of the Earth fell largely to the Arabs. The spread of Islam throughout the Near and Middle East, North Africa and parts of South East Asia enabled men to travel freely over this huge part of the globe and record their observations. Probably the best known of these travellers was Ibn Battuta, subject of Lucille McDonald's *The Arab Marco Polo* (Thomas Nelson, 1975). During the second quarter of the 14th century Ibn visited and described in detail almost every part of the Muslim world from southern Spain and West Africa, where among other things he described the then flourishing empire of Mali, to the lands of the Mongol Khans around the Yellow Sea.

With the renaissance the collection of geographical data spread to the west. The translation of Ptolemy's Geographia into Latin in 1405 was a turning point in the study of the Earth's surface. In calculating the circumference of the Earth Ptolemy arrived at a figure about one third too small and he also spread Eurasia over half the reduced circumference instead of over a third of it, (Fig.1). There have possibly been few errors that have proved more productive! Had it not been for this, Columbus might never have conceived the idea that Cathay would be reached by sailing westwards across the Atlantic - or even if he had, he might never have received the practical support from the King of Spain that made his expedition possible.

From the discovery of the New World in 1492 there began an era of geographical exploration and by the middle of the 18th century at least the broad outlines of all the major land masses on the Earth's surface had been etched in. Then between 1754 and 1765 the German philosopher Immanuel Kant worked on a philosophical basis for geographical study. The equation that he put forward and which can be read in his Gesammelte Schriften (Berlin Academy of Sciences, 1902-5), 'history : time :: geography : space' still remains as pertinent today as it was then. It certainly inspired Richard Hartshorne to write The Nature of Geography (Association of American Geographers, 1939), a book that remains probably the most searching discussion in English of the philosophy of geography.

In 1830 The Royal Geographical Society was founded in London and to mark the 150th anniversary of the event the society has published a work entitled Geography Yesterday and Today (O.U.P. 1980). The work is produced under the editorship of E.H. Brown and provides a useful review of the role of the RGS in the collection and analysis of geographic data from the beginning of the 19th century until the present day. It also contains valuable articles outlining the situation today in various branches of the discipline - such as biogeography, climate and the study of landforms, and historical and social geography. Each chapter contains useful references. Another recent publication that surveys the progress of the geographical sciences over this same period is T.W. Freeman's A History of Modern British Geography (Longman's, 1980). Following the body of the book are found short biographies of some 70 eminent 19th and 20th century British geographers such as Sir Halford MacKinder (1861-1947) remembered among other things for his Britain and the British Seas (1902) and Sir L. Dudley Stamp who is remembered more than anything for his organisation of the epic Land Utilisation Survey of Britain.

Today the final goal of the geographer is as it has always been - to conduct an ordered and systematic study of the Earth's surface and man's interaction with it. In the past the means of achieving this end were somewhat restricted, but today they are many and various. David Lanegran and Risa

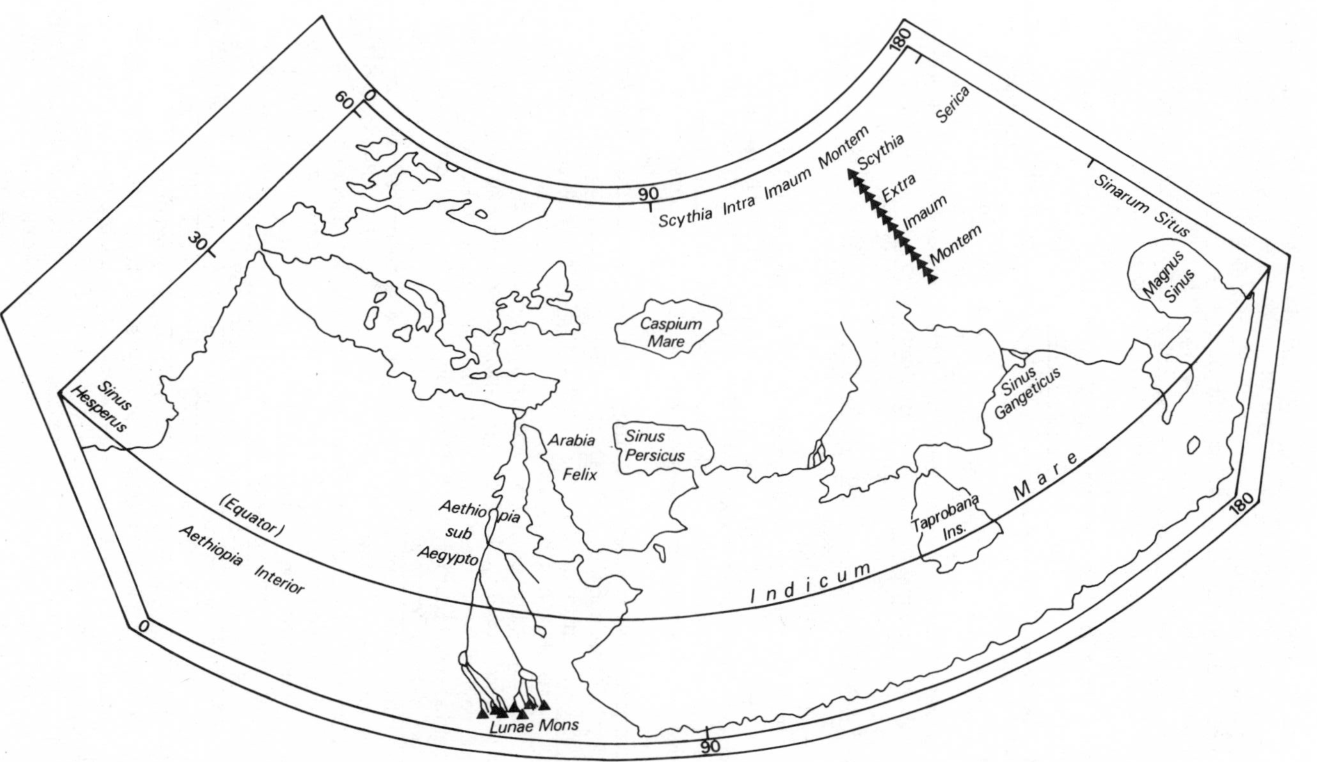


Figure 1. THE WORLD ACCORDING TO PTOLOMY

Palm's An Invitation to Geography (McGraw Hill, 1973) provides us with an interesting overview of modern geography. In it examples of many of the tenets of modern spatial analysis are given. Behaviour and radicalism in the geographical context are well illustrated by essays such as that on the geography of US riots and disorders in the 1960s,and a study of families with low incomes in rural America. The concept of diffusion is well illustrated by discussion of the spread of cholera in the US in the 19th century and of rock music in the 1950s. Unfortunately An Invitation to Geography is not well documented and therefore one has to go elsewhere to follow up many of the ideas discussed in the book. R.J. Johnsons's Geography and Geographers: anglo-American human geography since 1945 (Arnold, 1979) is a useful work to refer to in this context. To it are appended 30 pages of references. Other works that might prove useful are Location Analysis in Human Geography (Arnold, 2nd edn., 1977) by Peter Haggett, Andrew D. Cliff and Alan Frey and Fundamentals of Human Geography: a reader edited by John Blunden, Peter Haggett, Christopher Hamnett and Philip Sarre (Harper & Row, 1978).

Having stressed the importance of the spatial component in geography it is only fair to point out that there is a growing uneasiness today with the notion of the discipline as spatial analysis. Two books by Derek Gregory that have appeared recently and examine the nature of geography as we enter the final quarter of the 20th century are highly recommended to readers wishing to delve deeper into the philosophy of the subject. They are Ideology, Science and Human Geography (Hutchinson, 1978) and Social Theory and Spatial Structure (Hutchinson, 1980).

By the time this book comes into print more useful information and ideas will have been published and probably the best way of keeping abreast of this material will be to consult a serial publication that first appeared in 1969 entitled Progess in Geography (PIG). For the first seven years of its life this work appeared annually and was described as 'an international review of current research'. It endeavoured to outline work in the forefront of current developments in geographical research. 1977 saw two major developments in this serial publication. Firstly it was split into two separate publications; Progress in Human Geography: an international review of geography in the social sciences and the humanities and Progress in Physical Geography: an international review of geographical work in natural and environmental sciences. Secondly both these works were to appear quarterly.

An apparent dichotomy that may seem to exist in geographical studies is a division between a systematic and a regional approach. In the 19th and early part of the 20th century and particularly in France the regional approach was definitely in the ascendant. However today many geographers prefer to consider the spatial implications of specific phenomena rather than to consider the interaction of a wide range of phenomena giving particular character to a specific area of the Earth's surface. Fortunately some geographers still see their work having a regional significance and two recent works that should be consulted to show something of geography in its world-wide setting are Preston James One World Divided (Xerox 2nd edn., 1974) and H.J. de Blij's Geography: regions and concepts (Wiley, 2nd edn., 1978). In this context one should also keep an eye on Progress in Human Geography which periodically carries progress reports on particular areas of the world. A recent number of the work (Vol 4 (1980) pt.2) carried a short review of recent work concerning South East Asia.

## A Communication Network

Because a satisfactory bibliographic apparatus did not exist in classical Greece, Ptolemy's Geographia got lost for nearly 1500 years. Considering the impact it had when it was discovered it is impossible to surmise how world history might have been different had it not been lost to mankind in this way. For science of any kind to flourish and prosper the communication of ideas is vital. Today an elaborate network has been established to make every effort that all ideas, even if in the long run they prove to be as inaccurate as Ptolemy's were, are not lost to posterity. Progress in Geography discussed in the preceding paragraphs is just one facet of this modern network for bringing geographical ideas to an audience highly dispersed in time and space.

It is the object of this work to explain something of the apparatus for the communication of information as developed in differing fields of geographical science - hopefully to save the time and energy of persons delving for the first time into these branches of knowledge. There are two books that should initially be brought to the attention of persons wishing to come to grips with the literature of the geographical sciences. Firstly, Use of Social Sciences Literature (Butterworths, 1977) edited by Norman Roberts may be considered. In this book special attention is drawn to Roberts' introductory chapter which describes very succinctly the bibliographical system for the social sciences. Attention should also be drawn to J.L. Taylor's chapter on 'Environmental Planning Information'. The other useful title in the Butterworth's series is Use of Earth Sciences Literature (Butterworths, 1973) edited by D.H. Wood. In this volume readers are advised to refer to K.H. Clayton's chapter on 'Hydrology, glaciology, meteorology, oceanography and geomorphology' and W.A. Sarjeant and A.P. Harvey's section on 'Regional Geology'.

Before going any further with our analysis of the information sources for the geographical sciences two books must be mentioned that have played an important role in this field during the last decade; J.G. Brewer's The Literature of Geography: a guide to its organisation and use (Bingley, 1973) and

C.D. Harris' Bibliography of Geography Vol.1 (Univ. of Chicago, 1976). Both these works provide relevant information and much can be learned from them about the literature in this field.

However Brewer appeared half a decade ago and Harris over seven years ago and there have been a number of major developments - some for the better and some for the worse - in that time. We have already described major changes in the format and frequency of Progress in Geography. All these developments have occurred since both works were published. Both works point to the Royal Geographical Society's New Geographical Literature and Maps as an important bibliographical tool both for new books and maps. The publication listed additions to the Library and the Map Room of the Royal Geographical Society in London. Sadly the publication is no more - the last edition appeared in 1981. Brewer and Harris are also both the work basically of one compiler and it is felt to be better in a field like the geographical sciences to call upon the expertise of a team of contributors.

While both Brewer and Harris attempted to be comprehensive in their coverage of the subject this book makes no such claim. A Guide to Information Sources in the Geographical Sciences is a selective collection of essays by specialists in their particular branches of the subject. This introductory essay is followed by others on the information sources in four systematic subject areas - geomorphology, historical, agricultural and industrial geography written by practitioners in these respective fields. These four essays on systematic subjects are followed by four discussions of information sources on particular regions of the world. Chapters on Africa and South Asia provide us with information on major parts of the developing or under-developed world. They are followed by contributions on the U.S.A. and the U.S.S.R. - typical developed countries. Africa and South Asia are sometimes referred to as part of the Third World, as distinct from the free enterprise western states such as the U.S.A. or the state controlled economies of the Eastern Block of which the U.S.S.R. is the prime example. Again coverage is selective. Only sample areas are discussed by individual scholars familiar with the sources of information on the particular regions. The work is concluded with essays on four major tools for gleaning spatial information that may be used in any of the areas - both those discussed here and the many not specifically referred to. These tools are maps, aerial photos and satellite information, the computer and archival sources.

## Tracing the Relevant Book

Over the years the most important medium that has been used to record information and ideas has been the book. Today many new media are appearing, about which more will be said later in this chapter, but the book remains the first point in any basic literature search. Tracing the relevant book or books can be considered under two heads. Firstly one needs to know what has been published in the past on one's subject and a retrospective search is needed. Secondly it is important to know what is being published currently. Three great storehouses of information on geographical books published over the last century are Geographisches Jahrbuch, Bibliographie Géographique Internationale and the Research Catalogue of the American Geographical Society. Although the first two commenced publication in the 19th century, in 1866 and 1891 respectively, and a knowledge of their existence is important - there is little need in this context to examine them in detail. Geographisches Jahrbuch, while containing references to material in a variety of different languages,can not be most effectively used without a knowledge of German. Although it is still current (volume 65 appeared in 1979) its coverage is not now nearly as comprehensive as it was in the early decades of this century. Bibliographie Géographique Internationale does not present the same language problems as the Jahrbuch and its coverage is as good as ever it was. However neither of these works have appeared in a cumulative edition and therefore to make a thorough search in them involves looking in dozens of volumes.

For convenience of use the Research Catalogue of the American Geographical Society is way and ahead of the other publications. The Library of the American Geographical Society was founded in New York in 1923. From 1938 its accessions were listed in Current Geographical Publications,and in 1962 the cumulative holdings of the Library from its foundation until 1961 were published in 15 volumes by G.K. Hall in Boston (Mass). A decade later a first supplement to the Research Catalogue was published listing additions to the Library from 1962 to 1971 (Hall, 1972-4). The supplement appeared in four volumes. In 1978 a second supplement appeared with additions 1972-76. Thus one can scan many thousands of entries on a particular subject by simply looking in three sequences. One may compare scanning Bibliographie Géographique which would involve looking in something like 80 sequences or Jahrbuch that would involve searching in almost 70 different places. For publications after 1971 the Research Catalogue is kept up to date by Current Geographical Publications which appears ten times a year, monthly except for July and August. In 1978 the Library of the American Geographical Society was moved from New York to the University of Wisconsin at Milwaukee in Wisconsin. Current Geographical Publications is now therefore published by the University of Wisconsin.

The entries in the Research Catalogue and its updates are arranged in a classified order. The classification is largely a regionalone and only items that have little or no regional interest are arranged under systematic heads - described by the Americans as 'Topical' heads. This last point may be illustrated by observing that of the 15 volumes of the 1923-61 Catalogue only volumes 1 and 2 are devoted to 'Topical' entries - e.g. mathematical geography, physical geography, human geography, etc. Material in the remaining 13 volumes is arranged as follows:

V.3 The Americas as a whole, North America, Canada, U.S. (General)

V.4 U.S. North East, South East, North Central

V.5 U.S. South Central, West

V.6 Mexico and Central America, West Indies, South America (General), Columbia, Venezuela, Guianas

V.7 Ecuador, Peru, Bolivia, Brazil, Chile, Argentina, Paraguay, Uraguay and the Falkland Islands

V.8 Europe (General), British Isles, Low Countries, France

V.9 Switzerland, Germany, Scandinavia, European USSR

V.10 Poland, Czechoslovakia, Austria, Hungary, Rumania, Balkans, Italy, Portugal, Spain

V.11 Africa

V.12 Asia (General), Middle East, Soviet Central Asia, Tibet, Korea, Japan

V.13 China, South Asia, South East Asia

V.14 Australia, New Zealand and the Pacific Islands

V.15 Polar Regions, Oceans, Tropics

Entries for works on major regions are followed successively by entries for smaller areas within these regions. Thus in using the work it is advisable to consider material in larger regions or concerning broader topics, as works placed in such categories may well contain material on areas within them. It is interesting to observe that since the publication of the Catalogue in 1962 a much larger proportion of entries are being classified under systematic heads by the American Geographical Society. The first two volumes of the First Supplement (1962-71) published in 1972 contain regional entries arranged very much as in volumes 3-15 of the main Catalogue. The third and fourth volumes of the Supplement published in 1974 are devoted to systematic topics very much as volumes 1 and 2 of the 1962 publication. The same trend towards a larger number of systematic than regional entries appears to be continued in Current Publications in Geography. For example in the part of the serial publication for May 1979 (Vol.42 No.5) 45 pages are devoted to Topical, or systematic, 21 pages to regional entries.

This continuing bibliography provides a valuable tool for keeping in touch with new books in geography. However,while it will be found in many University and College libraries and larger Municipal libraries,a lot of readers might have difficulty finding a library that subscribes to it. The various national bibliographies that are published by all major book producing countries may be regarded as an alternative source of information on new geography books. National bibliographies are generally intended to comprise reference to all books produced within a given jurisdiction. Often the work will be compiled on the basis of books acquired by a National Library through the legal deposit of new books. Some national bibliographies also attempt to incorporate works published elsewhere but referring to their particular jurisdiction,and sometimes also books written by nationals but published elsewhere. The Kenya National Bibliography is following this pattern.

Two major national bibliographies are the British National Bibliography (B.N.B.) that has appeared weekly, quarterly and annually since 1950 and American Book Publishing Record (B.P.R.) that has appeared monthly and annually since 1960. Both publications list items by the Dewey Decimal classification scheme, BNB being based on copyright acquisitions at the British Library and BPR on American book production catalogued by the Library of Congress in Washington (DC). BPR is indexed by author and title while BNB carries author, title and subject indexes.

Keeping in Touch with Current Serial Publications

Books have quite a long gestation period - years can sometimes elapse between their conception and final appearance. More time must then be allowed for them to appear in the bibliographical record. The first appearance of new information and ideas will therefore usually take place on the pages of a serial publication rather than on those of a monograph.

Firstly it is useful to obtain some idea of the size of the periodical literature that is assisting in the communication of the geographical sciences. Probably the most comprehensive listing of titles has been compiled by C.D. Harris and J.D. Fellmann and is entitled an International List of Geographical Serials (Univ. of Chicago, 3rd edn., 1980). The work lists 3,445 serials from 107 different countries in 55 languages. It includes titles still current in 1980 and those that had been active in the field but had ceased publication by that date. Locations in some 30 union lists are also recorded. This work is supplemented by Professor Harris' companion volume, an Annotated World List of Selected Current Geographical Serials (Univ. of Chicago, 4th edn., 1980). This work lists 443 current geographical serials from 72 countries. A valuable feature of the Annotated World List....... is the annotation that accompanies each title and which does not appear in the International List..... The list of current cartographic Serials that follows chapter 10 illustrates the value of the annotations; they appear in the 'Brief Notes'. An Annotated World List... also contains a 48 page analysis of the citation of serials, both geographical and non-geographical, in geographical bibliographies.

Our knowledge of the periodical literature can be kept up to date by Ulrich's Quarterly (Bowker). Each issue lists some 2500 new publications which as above,are listed under 385 subject headings. Changes of title and cessations are also listed.

However serial publications usually contain numerous articles on different aspects of the subject. With monographs the title of the work alone can usually be a guide to the subject matter that is going to be found within the work. Here this is not the case and a more detailed analysis of the work's contents is required. In Ulrich's International Periodical Directory each of the subject heads is followed by a list of indexing and abstracting services that are active in the area. Under 'Geography' 19 such services are listed and 5 are referred to under the same head in Irregular Serials and Annuals. However one service from amongst these stands head and shoulders above the rest - this is Geo-abstracts which for the last ten years has been indexing and providing short abstracts of well over 10,000 items annually from several hundred serial titles.

In 1960 an annual abstracting service called Geomorphological Abstracts was founded. In its first year it abstracted 3,600 works. Five years later the work became known as Geographical Abstracts appearing in 4 sections. From 1972 it again changed its name, this time to Geo-abstracts,and it appeared in 6 parts. Finally in 1974 the publication began to appear in 7 parts and it has retained this format until today. The 7 parts each appear 6 times a year and they are as follows:

- A Land Forms and the Quaternary
- B Climatology and Hydrology
- C Economic Geography
- D Social and Historical Geography
- E Sedimentology
- F Regional and Community Planning
- G Remote Sensing, Photogrammetry and Cartography

An annual index to Geo-abstracts is now published regularly in 2 volumes. Volume one covers parts A,B,E and G, Volume 2 parts C,D and F. Cumulative indexes are also available for Geomorphological Abstracts (1960-65) in one volume, Geographical Abstracts (1966-70) in 4 volumes, Geo-abstracts Parts A-D (1971-75) again in 4 volumes, Parts E and F (1972-76) in two volumes and Part G (1974-8) in one volume.

Other Documents as Sources of Information

As well as books and serials there are various documentary forms in which information on the geographical sciences can be communicated. Government publications and the publications of international organisations such as the United Nations are one important category. Newspapers can also provide much relevant information and the relevance of university dissertations and conference proceedings speak for themselves. For guidance in these branches of the literature readers are directed to Guide to Reference Books (American Library Association, 9th edn., 1976),earlier editions of which were edited by Constance Winchell but which is now edited by Eugene Sheehy. A feature of this tool that is of special interest to the geographer is its broad international coverage. In its section on government publications for example a lengthy discourse on American government publications is followed by information guides to the government publications of nearly 50 other states. Included are listings of retrospective publications of a particular country or region and serial listings of current publications. An example of the former is Julian Witherall's French Speaking West Africa: a guide to official publications (Library of Congress, 1967). The work lists items from the mid-nineteenth century to the date of compilation appearing out of the eight independent Francophone states of West Africa - formerly colonies of France - and Togo,which from the end of World War I until independence was administered by France under mandate from the League of Nations and subsequently under trusteeship from the United Nations. An example of the second category is Canadian Government Publications/Publications du Gouvernement Canadien which has appeared annually since 1971.

Similarly with newspapers and dissertations the international coverage in Sheehy's Guide to Reference Books is very good. The newspapers of some twenty different states are documented. A fairly typical example is coverage of the press of India. The Annual Report of the Registrar of Newspapers which has appeared in New Delhi annually since 1956 provides an inventory of newspapers. The work provides a geographically arranged directory. Indian News Index indexes material from between 6 and 8 of the leading English language newspapers including the Times of India and Hindustan Times. Listings of the dissertations of 21 countries are provided. For most countries entries are arranged by subject and for those that are not, such as Dutch dissertations that are listed in Catalogus van Academische Geschriften in Nederland Verchenen by university, a subject index is usually provided. Dissertation Abstracts International (Univ. Microfilms, 1938- ) should also receive mention. Doctoral dissertations from several hundred universities in America and Europe are listed and a short abstract is given. Entries are arranged in a classified order and a copy of the microfilm edition of any of them can be obtained from University Microfilms,whose American base is Ann Arbor (Michigan) and whose UK office is in Bedford Row, London.

A somewhat similar tool to Sheehy is Walford's Concise Guide to Reference Material (Library Association, 1981). As well as dealing with the various major categories of reference material covered by Sheehy it also has a section entitled 'Area Study and Geography' that should prove of interest to readers of this book. Another valuable feature of Walford's Concise Guide is that it is able to refer to important works that have appeared in the 5 years since Sheehy was published. An example of this is South Asia Bibliography edited by J.D. Pearson (Harvester Press, 1979).

Four extremely important sources of information for the geographer in the study of the spatial organisation of the Earth's surface are maps, photographs (and particularly aerial photographs), statistical material and archives. However as chapters 10 to 13 are devoted to detailed discussion of these non-book or non-print materials the reader is referred to them for full particulars.

## Getting Hold of the Relevant Source Material

The first consideration in any effective communication system for the diffusion of scientific information and ideas is that the existence of the document or documents containing the relevant material must be brought to the attention of the potentially interested audience - albeit that this audience is highly dispersed geographically. A second consideration is that the interested party, having discovered the existence of relevant source material, should be able to physically get hold of it. It is not sufficient in most cases to simply know a book or periodical exists. Neither is it usually enough to read a short abstract. These simply whet the appetite and for maximum benefit the work needs to be physically consulted.

Libraries are the first step in fulfilling this need. Two of the best libraries for geographical source material in the English speaking world are those of the American Geographical Society, which since 1978 has been located in the Milwaukee Library of the University of Wisconsin and of the Royal Geographical Society in London. The record of the holdings of these libraries has been discussed earlier. In the U.K. another geographical library of note is that of the Geographical Association in Sheffield. For a major listing of important libraries and Universities in most of the countries of the world, World of Learning (Europa, annual) is an invaluable source. However, as well as academic and special libraries being able to bring you into physical contact with your source materials, municipal libraries will be able to satisfy many of the needs that a geographical scientist will present them with. Four particularly prestigious examples are those in Birmingham and Sheffield, England and the New York and Boston Public Libraries in the U.S.

However no library is going to be able to come up with everything. It has been suggested that a research worker in a leading British or American university will be very lucky if his library is able to come up with 70% of the material he needs in connection with his work! Consequently, libraries make every effort to co-operate with one another and Inter-Library Lending has grown to be an extremely important function of most libraries. In Britain, libraries in the main co-operate through British Library Lending Division (BLLD) based at Boston Spa in Yorkshire. And co-operation is not just a domestic thing. At the School of Oriental and African Studies of the University of London materials have been obtained through Inter-Library Lending from Peking! Similarly their material has on occasion been loaned to far flung corners of the world. In the U.S. co-operation operates similarly through the Research Libraries Group, whose headquarters are on the edge of the campus of the University of Chicago. Not surprisingly there is considerable liaison between Boston Spa and Chicago.

It can be a lengthy and costly business very often to transfer physically a book or periodical from one place to another, let alone the fact that many libraries have rules that prohibit rare and costly items leaving their premises. Today technology is playing an ever increasing role in helping to satisfy our information needs. Photocopying and microfilming are both very important. Few libraries are today without their batteries of photocopying machines and microfilm and microfiche readers. The computer also seems to have great potential. At the moment it is more successful in creating an awareness of a document's existence through reference to data bases such as GEO-REF and ENVIRON (Environmental Science Index) than actually bringing the document to the would-be reader. However, we can perhaps glimpse something of what the future holds in store for us with ever increasing use of the computer from Ifan Shepherd, Zena Cooper and David Walker's exciting little book Computer Assisted Learning in Geography (Council for Educational Technology with the Geographical Association, 1980). In this context we are obviously primarily concerned with the computer's role in retrieving and exploring stored bibliographic data. This book considers the computer's potential in this field alongside its ability to build and test models, map geographical information, play geographical games, demonstrate concepts of geographical form and to generally assist students of the subject with problem solving. This work, with its valuable listing of computer packages for the geographer and admirable bibliography is a must for any geographical library in the 1980's.

## The Universal Availability of Publications

Ideally in the search for information all published sources should be recorded to allow for universal awareness of their existence. And secondly these published sources should be made available universally. As has been mentioned in the previous paragraph, thanks to the computer the first of these considerations has come a long way to being met in the last decade or two. However relatively little progress has been made with the second. For several years the International Federation of Library Associations (IFLA) and UNESCO have had two objectives. Firstly they have been working towards Universal Bibliographic Control (UBC) - the bringing of everything published to the attention of the potential user. The International Standard Book Number (ISBN) and the International Standard Serial Number (ISSN) are all part of this concentrated internationl effort. This work's ISBN will be found on the back of the title page at the beginning of the book. The second objective is Universal Availability of Publications (UAP). To achieve this object the acquisition policies of libraries all over the world would have to be affected. Inter Library Lending too has to be perfected. Great strides are being taken by the international community in these directions - but there is still a long way to go. These factors aside there are still certain obstacles to Universal Availabiltiy that seem to present real problems. Finance is obviously a major consideration. Individual customs and censorship regulations have also to be considered. And last, but by no means least, there is the question of copyright. Different nations are requiring differing royalty payments for writers and are imposing differing rules relating to the copying of materials.

We still have a long way to go before we possess an ideal network for communicating man's spatial experience. But we have come a long way since it was possible to mislay Ptolemy's Geographia for 1500 years. Today, thanks to electronics and Universal Bibliographical Control, the modern Columbus would certainly know of the work's existence in a much shorter time. He might have to wait a while however to be 'misled' by its errors owing to financial stringencies at home, particularly severe censorship laws or copyright regulations. However thanks to the very real advances being made in the Universal Availability of Publications (UAP) it should not be too long before he can discover his New World.

THE SYSTEMATIC APPROACH

# Chapter Two

# GEOMORPHOLOGY

G.T. Warwick

On being approached by the editor to write this chapter my first thoughts were that it would be an easy task. A casual inspection of my own bookshelves and of the Geography and Geology sections of our university library soon disillusioned me and I realised what a growth industry I was examining, one neglected by the Guinness Book of Records!

Geomorphology is traditionally defined as the scientific study of the surface of the Earth and of the forces and agents that produced it. Often this excludes the form of the ocean bed, but this appears to me to be illogical. Most general texts then go on to distinguish between endogenetic and exogenetic forces: the former concerned with the building up of the surface above sea-level; and the latter being responsible for the destruction of the surface and its ultimate reduction to a smooth surface of gentle gradient. Latterly interest in the exogenetic forces has dominated geomorphological writing.

New concepts are rarely introduced via the medium of the book, though an increasing number of books consist of edited papers (sometimes with discussion) presented at special meetings. Other books collect papers which have had considerable impact on geomorphological thought, usually with a common theme. Several publishers have established series for this purpose, for example Dowden, Hutchinson and Ross with their important series, 'Benchmark Papers in Geology'. Several university presses also publish annual symposia on geomorphology such as that of the Press of the University of British Columbia which appears to be particularly active in this field.

In selecting works to include in this chapter, the following principles have been adopted. Firstly I have concentrated on books published since World War II, though admitting some from the war period and with some reference to classic works from earlier periods. Secondly I have concentrated upon books of use to university students and staff. It is surprising how quickly standard texts become outdated and require replacement or extensive revision. However it is rare today for a book to pass through more than three editions, although in the past this was certainly not the case. Sir Charles Lyell's Principles of Geology (Murray) for example passed through 11 editions between 1830 and 1872! This change is due in part to competition, associated with a larger market, but is mainly due to the explosive growth of knowledge.

## The History of Geomorphology

Geomorphology is rightly included in a book devoted to the geographical sciences, but it could equally well be included in one on the geological sciences and the reader is referred to Use of Earth Sciences Literature (Butterworths, 1973) edited by D.N. Wood. This useful work contains a contribution by the eminent geomorphologist Keith Clayton entitled, 'Hydrology, glaciology, meteorology, oceanography and geomorphology'. Parts of the subject would also be relevant to a source book on the engineering sciences. The name itself only dates from the turn of the century. Before that the subject was usually included within physical geology and physiography. Books on the latter were very similar in content to those on physical geography, in that they also dealt with climatology and biogeography. More recently physiography has come to be regarded as synonymous with geomorphology. The founding fathers of geomorphology were geologists and the history of the early period is well covered by G.L. Davies in The Earth in Decay. A History of British Geomorphology 1578 to 1878 (Macdonald, 1969). An even wider coverage is provided by R.J. Chorley, R.P. Beckinsale and A.J. Dunn in their multivolume History of the Study of Landforms (Methuen, 1964-). The first volume deals with the pre-Davisian era and the second with the life and work of William Morris Davis. There are two more volumes promised but they have yet to appear. A more limited period is covered by F.F. Cunningham in The Revolution in Landscape Science (University of British Columbia, 1966). The historical development of geomorphological ideas can also be traced in the papers edited by C.A.M. King in her volume Landforms and Geomorphology - Concepts and History. (Dowden, Hutchinson and Ross, 1976). James Hutton is usually taken to be the most influential of the early British writers on geomorphology and geology. It was he who introduced the Law of Uniformitarianism in The Theory of the Earth, with Proofs and Illustrations (Vols 1 and 2 Creech 1795). Some of the basic principles had been published ten years earlier in the Transactions of the Royal Society of Edinburgh in a paper which formed the first chapter of this book and which ended with the famous lines, '...we find no vestige of a beginning - no prospect of an end'. Volume 3 did not appear until 1899 and was published by the Geological Society of London. This delay was due to the manuscript having been mislaid. His drawings did not see the light of day until as recently as

1978 when they appeared as James Hutton's Theory of the Earth, the Lost Drawings (Scottish Academic Press), edited by G.Y. Craig with a text written by the editor assisted by D.Y. McIntyre and C.D. Waterson. The original volumes are now very rare and fetch high prices.

Physical Geography

Geomorphology was long taught as part of physical geography - often as the dominant part, with climatology and biogeography (especially vegetation) as adjuncts. Judging by the large number of college texts, this is still the case in the United States, though the sections devoted to biogeography usually include soils and they also deal with the extractive minerals. One of the most influential early texts was that of W.M. Davis and W.H. Sneyder, Physical Geography (Ginn, 1898) which was mainly devoted to the study of landforms. In Britain the most popular text was that of Philip Lake, Physical Geography (C.U.P., 1915) which passed through several impressions without revision until the second edition, revised by J.A. Steers, G. Manley and W.V. Lewis appeared in 1949. E. de Martonne's 3 volume Traité de Géographie Physique (Colin, 1925-7) also deserves mention here. It passed through no less than four editions in French and an abbreviated translation is available in English, translated by E.D. Laborde entitled A Shorter Physical Geography (Christophers, 2nd. edn., 1939).

A few texts included systematic physical and human geography in a single volume, such as V.C. Finch and G.T. Trewartha, Elements of Geography-Physical and Cultural (McGraw-Hill, 2nd edn. 1942), O.W. Freeman and H.F Raup, Essentials of Geography (McGraw-Hill, 1949) and in more recent times J.J. de Blij, The Earth- a Topical Geography (Wiley, 2nd edn, 1980). In Britain the introduction of an 'A' level syllabus in physical geography in the late 1940s and early 1950s led to a response from writers with works such as R.F. Peel's Physical Geography (English Universities Press, 1952) which concentrated upon geomorphology with a little geological background; the immensely successful The Principles of Physical Geography (University of London Press, 1954) by F.J. Monkhouse which went through numerous editions, and later A Concise Physical Geography (Hulton, 1972) ed.by D.Q. Bowen. Whilst there was little demand for a university text in physical geography in Britain at this time, in America the field was dominated by A.N. Strahler, later joined by his son A.A. Strahler. Early in the field came his Physical Geography (Wiley, 1951 and 1961), later published in a shortened form as Introduction to Physical Geography (Wiley, 4th edn. 1975). He also produced The Earth Sciences (Harper and Row, 1963) and with his son Environmental Geoscience (Hamilton, 1973), Introduction to Environmental Sciences (Hamilton, 1974), Geography and Man's Environment (Wiley, 1977) and Modern Physical Geography (Wiley, 1978). The latest batch of college texts, chiefly

American, is well represented by J.G. Navarra, Earth, Space and Time - an Introduction to Earth Sciences (Wiley, 1980) with a geological bias but also covering oceanography, meteorology and astronomy in non-technical language. Amongst others are A. Faniran and O. Ojo, Man's Physical Environment (Heinemann, 1980), W. Marsh and J. Dozier, Landscape - an Introduction to Physical Geography (Addison-Wesley, 1981) and D.K. Fellows, Our Environment - an Introduction to Physical Geography (Wiley, 2nd edn 1981), which are also in non-technical language.

During the last 15 years there has been a tendency for certain texts to integrate their subject matter, instead of producing water-tight compartments between the subjects. Amongst such books are W.B. Johnston (Ed.), Dynamic Relationships in Physical Geography (University of Canterbury, Christchurch, 1967) and R.J. Chorley and Barbara Kennedy, Physical Geography - a Systems Approach (Prentice-Hall, 1971). In this latter book the emphasis is on the systems approach and the manner in which it operates in the geomorphological, climatological and biological fields. The same principles are later extended to even wider fields by R.J. Bennet and R.J. Chorley in Environmental Systems (Methuen, 1978). A somewhat similar approach is made by G.H. Dury in An Introduction to Environmental Systems (Heinemann, 1981) but there is a greater emphasis here upon the environmental content. Some departments of Geography in British universities are now attempting to give integrated first year courses in physical geography, with emphasis upon the connections between the component subjects rather than their differences. In addition to the above texts F.E. Dohrs and L.M. Sommers (Eds.), Physical Geography - Selected Readings (Cromwell, 1967) is particularly useful in this context. At a more advanced level and with no professed attempt at integration, are the essays edited by R. Peel, M. Chisholm and P. Haggett in Processes in Physical and Human Geography (Heinemann, 1975). Professor Alan Straw also deals with the grouping as a whole in his inaugural address, Concerning Physical Geography (University of Exeter, 1972).

For reference there is a wide selection of dictionaries covering physical geography, environmental sciences and earth sciences, mainly published in the late 1970s. Amongst these are D. Dinely et al., Earth Resources: a Dictionary of Terms and Concepts (Arrow Books, 1976); A. Gilpin, Dictionary of Environmental Terms (Routledge and Kegan Paul, 1976); S.E. Stiegler (Ed), A Dictionary of Earth Sciences (Macmillan, 1976); M. Allanby, A Dictionary of Environmental Sciences (Macmillan, 1977); J.R. Holum, Topics and Terms in Environmental Problems (Wiley, 1977) and F.J. Monkhouse and J. Small, A Dictionary of the Natural Environment (Arnold, 1978). A more definitive work is McGraw-Hill's, Encyclopaedia of Environmental Science (1980). Practical work in physical geography has never been a popular subject, however A.A. Miller, The Skin of the Earth (Methuen, 1953) and B.J. Knapp, Practical Foundations of Physical Geography (Allen and Unwin, 1981) have tried to fill this niche.

In the last century many 'Physiographies' were produced for the intelligent layman, this field has been taken over by popular books on ecology. However R.S. Kandel, Earth and Cosmos (Pergamon, 1980) deals with the environment of man on Earth and with the environment of the Earth in the Cosmos.

## Physical Geology

The allied field of physical geology is dominantly concerned with geomorphological processes, but in addition generally includes an introduction to petrology, geological structures, vulcanicity and mountain building, and a discussion of the structure of the Earth. In American texts there is also a discussion of the occurence of economic minerals. In the U.S.A., C.R. Longwell, A. Knopf and R.F. Flint, A Textbook of Geology: Part 1 - Physical Geology (Wiley) was a pioneer in this field appearing in two editions before the Second World War, in 1937 and 1939. After the War the third edition in 1948 was published as a separate book. The work was later revised by a slightly different team of C.R. Longwell, R.F. Flint and J.E. Saunders and published under the same title by the same publishing house in 1969. The corresponding English book was A. Holmes, Principles of Physical Geology (Nelson), which first appeared in 1944 and was greatly expanded in 1965. It has been revised by his wife, D.L. Holmes, in a third edition published in 1978. She also adapted the text to form a shorter version entitled Elements of Physical Geology (Nelson, 1969). A. Holmes was a great teacher with a highly original mind, and his book contains much of his personal research and the concepts he helped to fashion. It is written in beautiful, clear prose which has inspired generations of post-war geology and geography students.

The 1970s witnessed a spate of American texts such as P.J. Wyllie, The Dynamic Earth. Textbook in Geoscience (Wiley, 1971), H. Lepp, Dynamic Earth: an Introduction to Earth Science (McGraw-Hill, 1973) and of course the indefatigable A.N. Strahler with his Principles of Physical Geology (Harper and Row, 1977). A new British book by R.A. Ridley, The Physical Environment (Ellis Horwood and Halstead, 1979) also deals with the atmosphere and the structure of the Earth along with the work of the agents of erosion. Quite recently Russian work in this field has been made available to English readers by the translation of G. Gorshkov and A. Yakushova, Physical Geology (Mir, 2nd edn 1977, based upon the third Russian edn of 1973). Most of the more recent textbooks are designed to cover introductory courses in geology, many of them for non-specialists, but few possess the originality which marked the pioneer texts.

## General Geomorphology Textbooks

Looking back, it is difficult to comprehend why it took so long for books devoted solely to geomorphology to appear on the market. Perhaps it was because the market was small and university texts could be produced to cover the whole of systematic geography or the whole of geology. Geography was dominated by regional geography and by modern standards the level of knowledge demanded was elementary. One of the earliest books in the field, which did not even use the name geomorphology in its title, was J.A. Marr, The Scientific Study of Scenery (Methuen) which ran through nine editions between 1900 and 1943. It was very much concerned with landforms, dealing successively with the structure of the Earth, climate, continents and oceans, mountains, valleys, lakes, volcanoes, plains and plateaus and deserts before considering ice as an agent of erosion and the work of the oceans in this context. In the British Commonwealth Sir Charles Cotton's Geomorphology - an Introduction to the Study of Landforms (Whitcombe and Tombs) was an early comprehensive text. It started its life as the Geomorphology of New Zealand, Part 1 - Systematic (Dominion Museum, Wellington, 1922) but Part 2 which was to consider regional implications was never written. The work only circulated widely in Britain in the post-war period in its 4th and 5th editions published in 1945 and 1949.

Another influential book of the inter-war era was J.A. Steers, The Unstable Earth (Methuen, 1932). It survived the War to appear in a 4th edition in 1945. A little later came S.W. Wooldridge and R.S. Morgan, The Physical Basis of Geography: an Outline of Geomorphology (Longmans, 1937) which followed the Davisian school, but was strongly influenced by many years of field work and teaching in south east England. It appeared in a second edition in 1959 with the title and sub-title reversed. This book and the teaching of its principal author had a widespread influence on British geomorphology throughout the 1940s and 1950s. However it was not universally accepted: I well remember as a young undergraduate being warned off reading it as 'the time is not yet ripe to start making generalisations in geomorphology'.

The slow development of geomorphology as a separate subdiscipline appears to have been paralleled in the United States, where the subject was generally regarded as being mainly geological and was well integrated with teaching in allied subjects such as sedimentology. One of the first general textbooks was that of A.K. Lobeck, Geomorphology - an Introduction to the Study of Landforms (McGraw-Hill, 1939). Strongly Davisian in organisation, it concentrated upon the study of landforms and their interpretation by the lavish use of photographs, diagrams and excerpts from topographical maps, while the text was reduced to a note-like form. A more prepackaged text of this period was P.G. Worcester, A Textbook of Geomorphology (Van Nostrand, 1939). A more individual book was O.D. Von Engeln, A Textbook of Geomorphology (Systematic and Regional) (Macmillan, New York, 1942), in which the author distilled a lifetime's research and wide reading. Another war-time issue was N.E.A. Hinds, Geomorphology - the Evolution of Landscape (Prentice-Hall, 1943), a massive and well illustrated tome. These pre-war and war-time books formed the background reading for British and American students until well after the War, for few new texts appeared in the immediate post-war period. In Belgium a notable exception was that of Paul Macar, Principes de Géomorphologie Etude des Formes du Terrain des Regions à Climat Humide (H.Vaillant-Carmanne, 1946), based on a course on geomorphology delivered in Oflag XC. In the following decade the most important textbook in English was W.D. Thornbury, Principles of Geomorphology (Wiley, 1954, 2nd edn 1969) which was organised on Davisian lines, but also covered some of the arguments of Penck and of American critics such as Kirk Bryan and his followers. It was written by a geologist for geologists and includes a thorough treatment of the role of geological structure and lithology in landform analysis. In France a similar function was exercised by M. Derruau, Précis de Géomorphologie (Masson, 1956) which had run to six editions by 1974.

In Germany a pre-war text by F. Machatschek, Geomorphologie (Teubner, 2nd edn 1935) was revised many times in the post-war period, and is now in its 11th edition, having been edited by H. Graul and C. Rathjens since the death of the writer in 1957. An English translation has been made by N.J. Davis, Geomorphology (Oliver and Boyd, 1969 from the 9th German edn 1968). Later it was rivalled by H. Louis, Allgemeine Geomorphologie (de Gruyter, 1960), now in its 4th edition with the assistance of K. Fischer. Stemming in part from this German tradition is the highly original Theoretical Geomorphology (Springer Verlag, 1960 and 2nd edn Allen and Unwin, 1970) by A.E. Scheidegger, which covers the exogenic processes and resulting landforms in terms of mathematical concepts and equations. Perhaps the most characteristic English book of this period was B.W. Sparks, Geomorphology (Longmans, 1966, 2nd edn 1972) which continued the Davisian-Wooldridge tradition. Later came A.L. Bloom's semi-popular book The Surface of the Earth (Prentice-Hall, 1969) and D.J. Easterbrook's Principles of Geomorphology (McGraw-Hill, 1969).

Competition between authors for this growing market has continued , A.F. Pitty, Introduction to Geomorphology (Methuen, 1971) broke out of the conventional mould by treating broad concepts, followed by the relationship between landforms and structure, the agents of erosion and their interrelation with landforms, and finishing with a treatment of landforms and time. This is more overtly geological in outlook than most geomorphological works produced by other geographers, and notably says little about climatic geomorphology. On the other hand H.F. Garner, The Origin of Landscapes (O.U.P., New

York, 1974) is more conventionally orientated towards climatic geomorphology, continually stressing the importance of climatic change and changes in base-level due to eustatic and tectonic causes. R.J. Small, The Study of Landforms - A Textbook of Geomorphology (C.U.P., 1970, 2nd edn. 1978) is still within the Davis-Wooldridge-Sparks tradition, its strong feature being the clarity of the writing and the simple, clear diagrams with which is is illustrated.

The emphasis of geomorphological work has long swung towards the study of process rather than landforms, and this is reflected in R.V. Ruhe, Geomorphology-Geomorphic Processes and Superficial Geology (Houghton Mifflin, 1975), which also pays more attention to superficial deposits than many other textbooks. K.W. Butzer, Geomorphology of the Earth (Harper and Row, 1976) affects a compromise between the older and newer views by first dealing with the agents of erosion and their effect upon structure, illustrating them by examples chosen from different climatic regimes, before finishing with a quick resumé of the morphology of the continents based largely upon geological considerations. R. Coque's Géomorphologie (Colin, 1977) starts with a background of geology and structure before dealing with agents of erosion, as does A.L. Bloom, Geomorphology, a Systematic Analysis of Cenozoic Landforms (Prentice-Hall, 1978). On the other hand R.J. Rice, Fundamentals of Geomorphology (Longmans, 1977) emphasises agents before dealing with structure on both a world and local scale. The emphasis upon the agents of erosion is most noticeable in two recent British works, E. Derbyshire, K.J. Gregory and J.R. Hails, Geomorphological Processes (Butterworths, 1979) and C. Embleton and J. Thornes (Eds.) Process in Geomorphology (Arnold, 1979). These two books bring together the results of much recent research work by geographers, geologists, civil engineers and oceanographers in a readable and easily digestible form. The former has a greater uniformity of style, whilst the latter brings together a wider body of expertise. It is also a sign of the growth of the subject that few can now claim to cover the whole field. There are signs that the fashion for agents of erosion has reached its peak, and there may be a return to landscape study, especially over time, and also a greater emphasis on the role played by geological structure and tectonics, with a regrowth of interest in a dynamic earth. The organisation of geomorphology in terms of climatic zones has also become less fashionable, but in certain specialised fields this is still of some importance to scholars.

Certain books popularising geomorphology, should also be mentioned. Of special note is Sir Arthur Trueman, Geology and Scenery of England and Wales (Gollancz, 1938 later revised and published by Penguin in 1972). Also important were Sir Dudley Stamp, Britain's Structure and Scenery (Collins, 1946) and G.H. Dury, The Face of the Earth (Penguin, 1959).

## The State of the Art

The general principles of geomorphology are most commonly treated in the introductory chapters of general textbooks and there are only a few books devoted to the setting out of these principles. Amongst these are P. Birot, Essai sur Quelques Problèmes de Géomorphologie Générales (Centro de Estudios Geograficos, Lisbon, 1949) and J. Tricart, Principes et Méthodes de la Géomorphologie (Masson, 1965). A more elementary treatment in English is P. McCullagh, Modern Concepts in Geomorphology (O.U.P., 1978).

There are however a few books, all edited volumes of lectures and essays, describing 'the state of the art'. One of these is C.A. Albritton (Ed.), The Fabric of Geology (Addison-Wesley, 1964), although it covers the whole of the subject there are at least three papers devoted to geomorphological topics. An unusual collection of papers was edited by G.H. Dury, Essays in Geomorphology (Heinemann, 1966) in which were published full-length technical essays on subjects in which the writers had established themselves as experts. Some of these essays are still relevant and are referred to in later writings. In what amounted to a Festschrift, E.H. Brown and R.S. Waters edited a volume entitled Progress in Geomorphology (Institute of British Geographers, 1974) consisting of essays written in honour of the late Professor David L. Linton. The two editors reviewed the progress in geomorphology since 1918, and then came a group of essays covering a wide range of subjects. A more deliberately orientated group of papers was included in C. Embleton, D. Brunsden and D.K.C. Jones (Eds.), Geomorphology, Present Problems and Future Prospects (O.U.P., 1978) which consisted of a series of papers presented to an international conference organised by the British Geomorphological Research Group, between 5-9 April, 1976 in London. Most of the papers were in the field of process geomorphology, or on the morphology of cold climatic regions and the tropics. The same group held another international symposium in March 1981, on the occasion of the 25th anniversary of the foundation of the Group, when methodological problems were discussed.

## Landforms

The avowed aim of most geomorphologists has been to describe and explain landforms. These two aspects form the main content of many textbooks of geomorphology, though some concentrate upon the processes of erosion at the expense of landform description. However a few authors have attempted to stick to description only. One such work is Vocabulaire Géographique, Tome II Le Relief (Documentation Géographique, 1958) which consists of a folder of annotated photographs of landforms mainly chosen from France. A more general application is W.F. Wood and J.B. Snell, A Quantitative System for Classifying Landforms (U.S. Department of the Army, 1960) and

M. Derruau, Les Formes de Relief (Masson, 2nd edn 1972). C.R. Twidale also produced a well illustrated volume on Analysis of Landforms (Wiley, Sydney, 1976) with many examples chosen from Australia.

Of individual landforms, planation surfaces have received most attention, usually in edited collections of reports, conference papers and collections of readings. An early post-war example was F. Ruellan et al (Eds.), Premier Rapport de la Commission pour l'Etude des Surfaces d'Applanissement autour de l'Atlantique (International Geographical Union, 1956). Others include M. Pecsi (Ed.), Problems of Relief Planation (Akadémie Kiadó, Hungary, 1970), G. Adams (Ed.) Planation Surfaces, Peneplains, Pediplains and Etchplains (Dowden, Hutchinson and Ross, 1975) and W.N. Melhorn and R.C. Flemal (Eds.), Theories of Landform Development (Allen and Unwin, 1980).

The study of soils or pedology, forms a separate branch of science from geomorphology and is often linked in geographical courses with biogeography. However the two subjects have many inter-connections which are brought together in A.J.W. Gerrard, Landforms and Soils (Allen and Unwin, 1981). This book is a valuable addition to the literature.

A closely allied topic to landform study is the description of the macro-relief of the continents and oceans on a world scale. Today few geomorphologists consider themselves competent to write such texts owing to the difficulty of keeping up with the literature. One brave soul who has attempted this was F. Matchatschek whose Das Relief der Erde (Gebruder Borntrager, 2nd edn 1955) covers this field in German. In English the main work is L.C. King, The Morphology of the Earth. A Synthesis of World Scenery (Oliver and Boyd, 2nd edn 1967). In France P. Birot accomplished an even greater task with his Les Régions Naturelles du Globe (Masson, 1970) which covers the relief and structure of the Globe plus an account of the climate, vegetation and soils of his geomorphological regions.

Mountain landforms have received a little attention. A popular account was given in the 'Que sais-je?' series by R. Fouet and C. Pomerol, Les Montagnes (P.U.F., 1955) and in Canada O. Slaymaker and H.J. McPherson (Eds.), Mountain Geomorphology: Geomorphological Processes in the Canadian Cordillera (University of British Columbia, 1966) covered this field. One of the latest works on mountain geomorphology which also deals with other aspects of the environment is L.W. Price, Mountains and Man: A Study of Process and Environment (University of California, 1981).

At the other end of the altitudinal spectrum is R.J. Russell, River Plains and Sea Coasts (University of California, 1967) which considers lowlands adjacent to rivers and the sea. It consists of four papers delivered as the Hitchcock Lectures for 1965.

Slopes

Landforms are made up of an assemblage of slope elements and since the last war there has been increasing attention paid to slope analysis and development. Much of the interest stems from the work of Walther Penck, Morphological Analysis of Landforms (Macmillan, 1953), originally published in German in 1924, but made available in English through the devoted translation of H. Czech and K.C. Boswell. However Penck's thesis that the surface morphology is the key to the interplay of tectonic and exogenic processes is not now generally accepted. Outside of the periodical literature one of the first publications to be devoted solely to slopes was P. Birot, P. Macar and J.P. Bakker (Eds.), Premier Rapport de la Commission pour l'Etude des Versants (International Geographical Union, 1956). Later came E.A. Yatsu, Rock Control in Geomorphology (Sozosha, 1966) calling attention to the importance of rock mechanics in the study of slopes. Also in 1966, the International Geographical Union held a symposium at Liège the proceedings of which were published as P. Macar (Ed.), L'Evolution des Versants (University of Liège, 1967). The results of a similar meeting in Britain were edited by D. Brunsden as Slopes, Form and Process (Institute of British Geographers, 1971). A collection of classic papers was reproduced by S.A. Schumm and P. Moseley (Eds.), Slope Morphology (Dowden, Hutchinson and Ross, 1973). The last decade saw the issuing of several books devoted entirely to slopes by individual authors. One of the most notable of these was M.A. Carson and M.J. Kirkby, Hillslope Form and Processes (C.U.P., 1972) which was particularly concerned with process. The same year also saw the publication of A. Young's Slopes (Oliver and Boyd). The same author joined with his wife to produce a slimmer volume designed for use in schools, A. Young and D.M. Young, Slope Development (Macmillan, 1974). The practical aspects of investigating slopes have been covered by A. Young, D. Brunsden and J.B. Thornes, Soil Profile Survey (Geo Abstracts, 1974) and by B. Finlayson and I. Statham, Hillslope Analysis (Butterworths, 1980). Studies have also appeared dealing with slopes on particular rock types, though many lithologies remain to be covered. They include A.L. Ward, The Stability of Keuper Marl Slopes (University of Birmingham, 1972) and M.J. Haigh, Evolution of Slopes on Artificial Landforms, Blaenavon, U.K. (University of Chicago, 1978).

Geomorphological Timescales

Recently there has been a resurgence of interest in the dating of landforms and superficial deposits, and in their rate of evolution. This return to older themes was marked by the appearance of J.B. Thornes and D. Brunsden's volume entitled, Geomorphology and Time (Methuen, 1977). Another book on this theme is R.A. Cullingford, D.A. Davidson and J. Lewis (Eds.), Timescales in Geomorphology (Wiley, 1980)., though it is chiefly concerned with Holocene events and deposits.

There has been a long history of co-operation between archaeologists and geologists, especially with those studying the palaeolithic period and its predecessors. In the post-war period such contacts have widened into the fields of geomorphology, climatology and palaeolimnology under the umbrella title of environmental archaeology. Parts of this literature are of more interest to geomorphologists than is generally realised, especially those sections dealing with artefacts found in superficial deposits in river terraces, raised beaches, etc.

## Process Geomorphology

The last 25 years have, however, witnessed a massive switch from the study of landforms and denudation chronology to process geomorphology. In geology too there has been a similar development, especially in North America, as exemplified by A.M. Johnson, *Physical Processes in Geology* (Freeman, Cooper and Blackwell, 1970). A more purely geomorphological book is M.A. Carson, *The Mechanics of Erosion* (Pion, 1971), which applies the principles of soil and rock mechanics, as well as other physiographic processes. Similar work in the Low Countries was reported in P.W. Birkeland (Ed.), *Pedology, Weathering and Geomorphological Research* (O.U.P., New York, 1976). A more specialised theme is discussed in *Thresholds in Geomorphology* (Allen and Unwin, 1980) edited by D.R. Coates and J.D. Vitek. Attention will be concentrated upon fluvial and marine processes. The effects of ice and wind are also important in particular parts of the world and the reader is directed for further references in these areas to the general texts quoted above.

## Fluvial Processes

Rivers and associated overland flow have always been central to the study of most landscapes, but there were few books which concentrated on river action until about twenty years ago. Probably the most influential book was L.B. Leopold, M.G. Wolman and J.P. Miller's *Fluvial Processes in Geomorphology* (Freeman, 1964) which summarised much of the theoretical and practical work of the previous thirty years. In France M. Pardé had summarised knowledge in this field in his *Fleuves et Rivières* (Colin, 3rd edn., 1955). At a somewhat more elementary level was M. Morisawa, *Streams, their Dynamics and Morphology* (McGraw-Hill, 1968). Further general texts were provided by S.A. Schumm, *The Fluvial System* (Wiley, 1977) and M. Morisawa, *Rivers* (Longmans, 1981).

The publication of collections of classic papers is a sign of the coming of age of a topic, and in this case the market has been provided with G.H. Dury (Ed.) *Rivers and*

River Terraces (Macmillan, 1970) and S.A. Schumm (Ed.), River Morphology (Dowden, Hutchinson and Ross, 1972). Symposia on fluvial processes have also flourished, amongst the more recent examples being the International Association of Scientific Hydrology (IASH), which since 1971 has been known as the International Association of Hydrological Sciences (IAHS), Symposium on River Hydrology,Berne, 1967 (IASH, 1968); M. Morisawa (Ed.), Fluvial Geomorphology (Allen and Unwin, 1973) and R. Davidson-Arnott and W. Nickling (Eds.) Research in Fluvial Geomorphology (Geo Abstracts, 1978). The most recent work in this area is Adjustments of the Fluvial System (Allen and Unwin, 1981) edited by D.D. Rhodes and E.J. Williams. In addition to these collected works there have been two useful collections of essays, A. Pitty (Ed.) Geographical Approaches to Fluvial Processes (Geo Abstracts, 1979) and J. Lewin (Ed.), British Rivers (Allen and Unwin, 1981). Whilst the latter dealt specifically with British examples, it covered a wide range of fluvial geomorphology.

There are also many books which do not cover the whole range of fluvial processes, though they are not confined to a single topic; several of which embrace other branches of knowledge not treated in this chapter. A good example is R.T. Ogilby et al (Eds.), River Ecology and Man (Academic Press, 1972) which covers biological aspects as well as river morphology and geomorphological processes. Contributions from engineers under this heading include H.W. Shen (Ed.), River Mechanics (Shen, Fort Collins, Colorado, 1971) and H.W. Shen (Ed.), The Modelling of Rivers (Wiley, 1974). Of more specialised interest is D. Kimball, Denudation Chronology. The Dynamics of River Action (Institute of Archaeology, University of London, 1948), M.J. Kirkby, The Stream Head as a Significant Geomorphic Threshhold (Department of Geography, University of Leeds, 1978) and L. Starkel and J.B. Thornes, Palaeohydrology of River Zones (Geo Abstracts, 1981). There are also three useful reference books which cover parts of the field of riverine processes. They are R.K. Greswell and A. Huxley (Eds.), Standard Encyclopaedia of the Rivers of the World (Weidenfeld and Nicholson, 1965) which provides standardised descriptions of the major rivers, with a bias towards British streams; D.K. Todd (Ed.), Water Encyclopaedia (Water Information Center, New York, 1970) and E. Vollmer (Ed.), Encyclopaedia of Hydraulics, Soil and Foundation Engineering (Elsevier, 1967) which is a glossary of terms used in the subjects covered.

## Fluvial Processes - Hydrology

Hydrology is an applied science dealing with all aspects of water in all its physical states. In practice it is mainly concerned with water supplies from rivers, lakes (including reservoirs) and ground water. Hydrologists pay considerable attention to the volume of water in river channels, its velocity and variations in discharge. It is a science which embraces parts of climatology, civil engineering, geology, geomorphology, economics, biology, etc.

During the last war and sometime afterwards the standard hydrological reference work was O.E. Meinzer (Ed.) Hydrology (McGraw-Hill, 1942, facsimile reprint, Dover, 1949), which covered most of the physical aspects, including two useful chapters on the hydrology of limestone and volcanic terrains. A later, more popular text was P.H. Kuenen, Realms of Water (Cleaver-Hume and Wiley, 1955), translated from the original Dutch by M. Hollander. The last twenty years or so has seen a steady stream of texts such as C.O. Wisler and E.F. Brater, Hydrology (Wiley, 2nd. edn., 1959), M. Roche, Hydrologie de Surface (Gauthier Villar,1963) and A. Guilcher, Précis de Hydrologie (Masson, 1965) which dealt with oceans and lakes as well as rivers. R. Furon's The Problem of Water - A World Study (Faber and Faber, 1967) was translated from the French text of 1963 by H. Barnes and ranges widely over the hydrological field. An interesting development was a synthesis of hydrology, geomorphology and socio-economic geography edited by R.J. Chorley, Water, Earth and Man (Newnes-Butterworth) which was subsequently published separately in three parts entitled Physical Hydrology, Introduction to Fluvial Processes and Introduction to Geographical Hydrology. As in other branches of geomorphology the 1970s saw a spate of texts such as P.S. Eagleson, Dynamic Hydrology (McGraw-Hill, 1970), W.C. Walton, The World of Water (Weidenfeld and Nicholson, 1970), J.L. Astier, Géophysique Appliquée a Hydrologie (Masson, 1971) and R.A. Deju, Regional Hydrology Fundamentals (Gordon and Breach, 1971). Later came R.G. Kazmann, Modern Hydrology (Harper and Row, 2nd edn.,1972) and R.C. Ward, Principles of Hydrology (McGraw-Hill, 2nd edn.,1975). A new theoretical viewpoint was presented by D.H. Miller, Water at the Surface of the Earth. An Introduction to Ecosystem Hydrodynamics (Academic Press, 1978) which dealt with water mass, material carried by water and the energy transformations involved, by budget accounting methods. A reference book providing data on the major hydrological and ecological features of the rivers of the U.S. was R.M. Cushman, S.B. Bough, M.S. Morin and R.B. Craig, Source Book of Hydrology and Ecological Features - Water Resources Regions of the Conterminous United States (Ann Arbor Science, 1980).

There have also been published a series of volumes devoted to the proceedings of hydrological conferences, and collections of essays on similar themes. One of the latter was J. Dresch et al (Eds.), Hydrologie-Mélanges Offert par ses Amis et Disciples à Maurice Pardé (Ophyris, 1969), which covered a wide hydrological range but specialised on floods, of which Pardé had made a special study. There have been at least three national symposia for professors of hydrology under the general title The Progress of Hydrology, organised and published by the U.S. National Science Foundation between 1969 and 1971. There have also been international hydrological congresses, of which the third was held at Colorado State University in 1977 and the proceedings of which were published in H.J. Morel and others as Surface and Subsurface Hydrology (Water Resource Publications, 1979).

Fluvial Processes - Underground Water

Some authorities use the term underground water to include all water below the surface, dividing up into soil water, vadose water (above the water table) and phreatic water or groundwater in the saturated zone. Soil water is perhaps best dealt with in Soil Science, which is outside the scope of this chapter. However soil water is important to geomorphological processes and there have been a few books published on it. One of the most comprehensive is E.C. Child's, An Introduction to the Physical Basis of Soil Water Phenomena (Wiley, 1969). The topic was also discussed by several authors in M.J. Kirkby (Ed.), Hillslope Hydrology (Wiley, 1978). A more specialised aspect that has recently engaged the attention of geomorphologists is soil piping, the development of tube-like channels in the soil horizons. It has been dealt with by J. Gilman and M.D. Newson in Soil Pipes and Pipeflow (Geo Abstracts, 1980) and again by J.A.A. Jones, Soil-piping: A Literature Review (Geo Abstracts, 1980).

Groundwater is primarily within the province of the hydrogeologist, but since it plays such an important part as a reservoir for the hydrological cycle on land and also acts as a solute of mineral matter from rocks it is necessary for the geomorphologist to be aware of its properties and mode of occurrence. Before the war the classic book on this subject was C.F. Tolman, Ground Water (McGraw-Hill, 1937). It has been succeeded amongst others by C.F. Todd, Groundwater Hydrology (Wiley, 1959, 2nd. edn., 1980) and R.J.M. de Wiest, Geohydrology (Wiley, 1965). Much of the information about groundwater comes from the drilling of wells, the subject of E.E. Johnson's Groundwater and Wells (Johnson, 1966). Since wells are expensive to drill, hydrogeologists have developed mathematical models to enable them to predict what is happening between well sites and this is the subject of P.A.D. Domenico, Concepts and Models in Groundwater Hydrology (McGraw-Hill, 1972). A more indirect method of interpretation is given by K.E. Nefedov and T.A. Popova, Deciphering of Groundwater from Aerial Photographs (Amerind Pub., New Delhi, 1972). A guide to foreign hydrological literature was provided by H.O. Pfannkuch, Elsevier's Dictionary of Hydrology in three languages (English, French and German) (Elsevier, 1969).

Of special interest to geomorphologists is the hydrogeology of limestone terrains and to a lesser extent of igneous rocks. Both topics were covered by the IASH Symposium, Hydrology of Fractured Rocks, 2 vols., (IASH, 1968), Dubrovnik,

1967, but only one communication dealt with igneous rocks. Dubrovnik was also the venue for a US-Yugoslav symposium, the proceedings of which were published by V. Yevyevich (Ed.), Karst Hydrology and Water Resources (Water Res. Publns, 1976) in an English edition. A German text on this topic was also recently made available in English, A. Bögli (Translated by J.C. Schmidt), Karst Hydrology and Physical Speleology (Springer Verlag, 1980). Most books on physical speleology and karstic morphology also include material on hydrology.

Fluvial Processes - Drainage Basin Studies

The drainage basin is a fundamental unit for both the hydrologist and the geomorphologist and there is a great deal of overlap in their work and no attempt will be made to separate them. Studies of river basins or watersheds are concerned with the transfer of masses of water, sediment and solutes through them. The measurement of discharge at selected points is fundamental for most of such work which is often published in the scientific journals or remains buried in research theses. However there have also been a number of special conferences on instrumented river basins, especially during and after the hydrological decade, as well as reports from single authors.

The International Association of Scientific Hydrology (IASH) has played a leading role, amongst the reports of symposia organised by them were Representative and Experimental Areas in 2 vols, (IASH, 1965), Results of Research on Representative and Experimental Basins (IAHS, 1973) and again The Hydrological Characteristics of River Basins and the Effects of those Characteristics on Better Water Management (IAHS/Scientific Council of Japan, 1975). A national meeting on similar lines resulted in C.M. Moore and C.W. Morgan (Eds.), Effect of Watershed Changes on Stream Flow (Univ. of Texas Press, 1969).

A guide to individual studies was published by the American Geophysical Union, An Inventory of Representative and Experimental Watershed Studies Conducted in the United States (AGU, 1975). Examples of such work in the U.S.A. included M. Morisawa, Relationship of Quantitative Geomorphology to Stream Flow in Representative Watersheds of the Appalachian Plateau Province (ONR, 1959) and I.D. Correll (Ed.), Watershed Research in Eastern North America (Smithsonian Institute, 1977). Other examples may be found amongst the United States Geological Survey Professional and Water Supply Papers (referred to below as USGS). From Britain a collection of similar work was published as, K.J. Gregory and D.E. Walling, Fluvial Processes in Instrumented Watersheds (Institute of British Geographers, 1974). Another publication in the English language was A.C. Imeson and H.J.M. Van Zon, Erosion Processes in Small Forested Catchments in Luxembourg, (Geo Books, 1979). An allied theme was the subject of P. Beaumont's River Regimes in Iran (Dept. of Geog, Univ. of Durham, 1973) and of I. Douglas's Rates of Denudation in Selected Small Catchments in Eastern Queensland (Univ. of Hull, Dept. of Geog., 1973).

Two books which brought together the general findings from many individual studies were K.J. Gregory and D.E. Walling, Drainage Basin Form and Process (Arnold, 1973) and D.I. Smith and P. Stopp, The River Basin - An Introduction to the Study of Hydrology (C.U.P., 1978).

## Fluvial Processes - Run-off

Geomorphologists become interested in water when it arrives, as raindrops, upon the surface of the land. Some of this rainwater causes erosion, some sinks into the soil and the remainder runs off or evaporates. This phase of the hydrological cycle forms the subject matter of books by E.E. Foster, Rainfall and Runoff (Macmillan, New York, 1948); N.P. Chebotarev, Theory of Stream Runoff (Israel Program for Scientific Translations, 1966) and W.W. Emmett, The Hydraulics of Overland Flow on Hillslopes (USGS, 1970). Also much of M.J. Kirkby (1978 -see above) embraced this topic. D.R. Weyman, Run-off Processes and Streamflow Modelling (O.U.P., 1975) provided an easily digested introduction. Towns produce special problems with their hard road and roof surfaces, encouraging rapid run-off, and this was the subject of an international symposium whose proceedings were edited by M.B. McPherson under the title, Research in Urban Hydrology (UNESCO, 1977). The concentration of surface flow on steep, degraded arable slopes may lead to the formation of rills and gullies, but that is a field of its own and is generally treated under soil science and as such lies outside the scope of this chapter.

A special form of run-off is associated with glaciers and ice-sheets, but is generally treated under the heading of fluvio-glacial or glaciofluvial action in books dealing with glaciers and glaciation. However there was one symposium organised by the Glaciological Society and IASH at Cambridge in 1969 which dealt purely with the hydrological aspects, J.W. Glen, R.J. Adie and D.M. Johnson (Eds.), Symposium on the Hydrology of Glaciers (IAHS, 1973).

## Fluvial Processes - The Achievement of a Steady State

Many rivers exhibit sections of their courses where their longitudinal profiles form smooth logarithmic curves and on balance are neither excavating their beds nor consistently depositing debris. Thus over a period of years they retain the same general characteristics having achieved a steady state. This has been a subject of discussion by geomorphologists for over 150 years. The general position has recently been reviewed by E.A. West, The Equilibrium of Natural Streams, (Geo Books, 1978).

Fluvial Processes - Stream Channels

Rivers flow in confined channels for most of their time, though bank-full stage has been found to be most important in determining the characteristic depth and width of channels. Relationships have been discovered between these properties and the discharge and mean velocity of streams. Similarly there has been a great deal of emphasis on the relationships between the different segments of a drainage network, the density of streams and their rate of coalescence. In addition engineers have made a special study of flow in both natural and artificial channels in order to predict the effect of increasing discharge and to design river control works.

One of the best geomorphological texts was that of Leopold, Wolman and Miller (1964) - which together with some of the later work on rivers cited above, paid considerable attention to river channels. A slimmer and more recent book was J. Thornes, River Channels (Macmillan, 1979). Another, more specialised publication, was K.J. Gregory (Ed.) River Channel Changes (Wiley, 1977).

Another group of publications deals with the interaction of water with the sides and bed of channels. These include I.L. Rozovoskii, Flow of Water in Beds in Open Channels (Israel Program for Scientific Translations, 1961) and D.B. Simons and E.V. Richardson, Resistance to Flow in Alluvial Channels (USGS, 1966). The roughness of natural channels has an important effect upon the velocity and discharge of water flowing in them and is an important factor in calculating the velocity of flow. An important guide to make assessments of this was provided by H.H. Barnes, Roughness Characteristics of Natural Channels (USGS, 1967). A more specialised work in this field was J.P. Miller, High Mountain Streams; Effects of Geology on Channel Characteristics and Bed Material (New Mexico State Bur. of Mines and Min. Res., 1958).

The inter-relation of channel depth, width and stream velocity and discharge, known as hydraulic geometry, was of continuing interest after the publication of L.B. Leopold and T. Maddock, The Hydraulic Geometry of Streams and some Physiographic Implications (USGS, 1953). A later offshoot of this work was J. Thornes, Speculations on the Behaviour of Stream Channel Width (London School of Economics, Dept. of Geography, 1974).

Stream channels may adopt linear, meandering or braided courses and once again one finds the geomorphologists of the US Geological Survey giving a lead, together with collaborators whose work they published. L.B. Leopold and M.G. Wolman dealt with the general problem in their much-quoted River Channel Patterns - Braided, Meandering and Straight (USGS, 1957). In the previous year L.B. Leopold and J.P. Miller covered the allied topic of Ephemeral Streams - Hydraulic Factors and their Relation to the Drainage Network (USGS, 1956). Meanders have attracted considerable attention from geomorphologists and engineers, with notable contributions from R.A. Bagnold, Some Aspects of the Shape of River Meanders (USGS, 1960); W.B. Langbein and L.B. Leopold, River Meanders - Theory of Minimum Variance (USGS, 1966) and more recently S. Martval and G. Nilsonn, Experimental Studies of Meandering (UNGI, 1972).

Fluvial Processes - Flooding

In times of excessive discharge, water flow is no longer confined to river channels, which 'burst their banks' and inundate the adjacent flood plains. On such occasions fast-moving flood water can modify the bed and sides of river channels and may even change the course of the river. The damage may include property, roads and bridges so that there is much encouragement for engineers to construct control works to minimise such damage. There is in consequence a large body of literature upon this subject. One of the earliest works in our period was W.S. Hoyt and W.B. Langbein, Floods (Princeton Univ. Press, 1955) and a recent comprehensive book was R.C. Ward, Floods. A Geographical Perspective (Macmillan, 1978). From the engineering side came Tennessee Valley Authority, Floods and Flood Control, (TVA, 1961) and the Institution of Civil Engineers, River Flood Hydrology (ICE, 1965) Floods after particularly intensive rainstorms are specially dangerous with very quick responses in the rivers, and the IAHS held a special meeting on this, the results of which were entitled Flash Floods (IAHS/WMO, 1974).

Fluvial Processes - Sediment and its Transport

Streams transport matter in suspension, in saltation, or as bedload, depending upon the size and density of the particles available. One of the most important studies ever made in this field was G.K. Gilbert, The Transportation of Debris by Running Water (USGS, 1914) which is still quoted by most writers on this subject, and was the source of much current terminology. Once again R.A. Bagnold made a significant contribution with An Approach to Sediment Transport from General Physics (USGS, 1966). Other recent books include W.H. Graf, Hydraulics of Sediment Transport (Academic Press, 1971); J. Bogárdi, Sediment Transport in Alluvial Streams (Akédemiai Kiadó, Budapest, 1974) and M.S. Yalin, Mechanics of Sediment Transport, (Pergamon, 1977).

Besides transporting solid material streams also carry considerable amounts of dissolved chemicals in solution. Much of the relevant literature is concerned with the techniques of chemical analysis, but the general principles are covered by J.D. Hem, Study and Interpretation of the Chemical Characteristics of Natural Water, (USGS, 1970).

Fluvial Processes - Deposition

When the velocity and turbulence of streams falls below critical threshold values, then deposition occurs in the stream bed, though with changed conditions, this material may be taken up again. Flood deposits however may remain longer, after a flood has subsided, producing floodplains and levées, etc. Engineers are interested in controlling deposition, especially in navigable waters, whilst geologists are interested in the nature and processes of fluviatile sedimentation, as part of the much wider field of sedimentary petrology.

An important contribution to the study of riverine and other sediments is G.V. Middleton (Ed.) Primary Sedimentary Structures and their Hydrodynamic Interpretation (Soc. of Economic Palaeontologists and Mineralogists, 1965). A more specifically river-oriented conference was held in the University of Calgary, the proceedings being edited by A.D. Miall, entitled, Fluvial Sedimentology, Proceedings of a Symposium held 20-22 October 1977, (Canadian Soc. of Petroleum Geologists, 1978). The effects of alluviation may also be studied in the laboratory with the aid of large-scale models, as was reported by J.F. Friekin, A Laboratory Study of Alluvial Rivers (US Corps of Eng., Vicksburg, 1945). Allied to floodplain deposits are those of alluvial fans, especially noticeable in the valleys of glaciated mountains, and a recent study of these land forms has recently appeared, by A. Racchi, Alluvial Fans: an Attempt at an Empirical Approach (Wiley, 1981). Within the channel itself are many minor constructive landforms, such as point bars, dunes, antidunes, etc, and W.B. Langbein and L.B. Leopold made a special study of some of these in River Channel Bars and Dunes - Theory of Kinematic Waves (USGS, 1968).

River deltas have received rather more attention due to their rapidly changing features and wide variety of landforms and depositional structures. The IASH held a special symposium in Bucharest in May 1969 on The Hydrology of Deltas, 2 Vols, (IASH, 1970), which dealt mainly though not exclusively with eastern European deltas. Because deltas are commonly found to be associated with oil and gas fields, geologists have taken a special interest in their structures. Amongst the many conferences on this aspect may be cited, W.L. Fisher, et al., Delta Systems in the Exploration for Oil and Gas (Bureau of Econ. Geol., Univ. of Texas, 1969); J.P. Morgan and R.H. Shaver (Eds.) Deltaic Sediments, Modern and Ancient (Soc. Econ. Pal. & Mineral. 1970) and M. L. Broussard (Ed.), Deltas, Models for Exploration, (Houston Geol. Soc., 1975).

Fluvial Processes - Conclusion

The importance of fluvial processes is obvious from the large literature devoted to them. This is largely due to the importance of rivers to man. However the costs of many detailed observations are high, but there are signs that all concerned are beginning to co-operate and discover that each group of research workers has something to offer. It is certain that the future will see more multi-disciplinary conferences on specific themes within this vast field of study.

Marine Processes

The sea acts as a base-level for rivers, but it is a very active agent in its own right, chiefly along the coastlines of the land-masses. It works chiefly through wave-action, whose effects are increased by tidal variations in sea level. It produces its own special landforms, the more permanent ones etched into the land as marine-erosion platforms, often backed by cliffs. Depositional forms occur as beaches, spits and offshore bars or barrier beaches and islands. In some places the coast is backed by one or more lines of dunes. In plan most coastlines consist of a series of embayments separated by headlands and interrupted by estuaries where large rivers discharge into the sea. Within the tropics there are special features such as coral islands, atolls and reefs.

The field is well served with books. In 1959 C.A.M. King published her Beaches and Coasts (Arnold, 2nd edn. 1972) which has remained a standard British reference work. Another important book was V.P. Zenkovich (ed. by J.A. Steers) Processes of Coastal Development (Oliver and Boyd, 1967), translated from the Russian. C. Putnam et al. (Eds.) initiated a global look at coasts with Natural Coastal Environments of the World (University of California Press, 1960) which was further developed by J.L. Davies, Geographical Variation in Coastal Development (Longmans, 2nd edn., 1980). One of the latest books was R.E. Snead, Coastal Landforms and Surface Features. A Photographic Atlas and Glossary (Hutchinson and Ross, 1981). There are at least three encyclopaedias for reference in this field; R.W. Fairbridge, (Ed.), The Encyclopaedia of Oceanography (Hutchinson and Ross, 1966), the McGraw-Hill Encyclopaedia of Ocean and Atmospheric Sciences (McGraw-Hill, 1979) and G.C. Groves and L.M. Hunt, The Ocean World Encyclopaedia (McGraw-Hill, 1980).

Oceanography was included by many geographers as part of physical geography, but it has expanded into a separate subject, often divided into physical and biological oceanography, of which the former is most relevant to the geomorphologist. Physical oceanography includes the study of sea-level and its changes, tides, all types of marine waves, currents, the physical and chemical properties of sea-water and often the form of the sea-floor and marine sediments. These last two also constitute the main contents of books on marine geology and geomorphology.

There are a number of all-embracing oceanographic books, only part of whose contents are of direct interest. These include C.A.M. King, Introduction to Physical and Biological Oceanography 2 vols. (Arnold, 2nd edn., 1975); C. Drake, et al., Oceanography, (Holt, Rinehart and Whiston,1978); H.W. Menard, Ocean Science (Prentice-Hall, 1979) and C. Ditrich and H.J. Roll (Trans. by S. Roll), General Oceanography (Wiley, 1980 , based on the 3rd German edn. of 1975), to quote some of the more recent issues. Again there are collections of classic papers such as M.B. Deacon (Ed.), Oceanography, Concepts and History (Dowden, Hutchinson and Ross,1978) and symposia, e.g., H. Charnock and Sir George Deacon (Eds.), Advances in Oceanography (Plenum Press, 1978).

In the more restricted field of physical oceanography, recent texts include J.A. Knauss, Introduction to Physical Oceanography (Prentice-Hall, 1978) and G.L. Pickard, Descriptive Physical Oceanography (Pergamon, 3rd edn., 1978). An encyclopaedic series was edited by E.G. Goldberg entitled The Sea - Ideas and Observations in Progress in the Study of the Sea, 5 vols. (Wiley, var. dates), dealing with different aspects of physical oceanography, each volume being produced by a separate editor. There was also a useful collection of essays edited by A.L. Gordon, Studies in Physical Oceanography A Tribute to Georg Wüst on his 80th Birthday, 2 vols, (Gordon and Breach, 1972).

### Marine Processes - Sea Levels and Tides

The sea acts as a base-level towards which rivers are graded, but the level is always changing, both diurnally and over periods of geological time, especially during the Quaternary era. Books on sea level changes have been written by E. Lisitzin, Sea Level Changes (Elsevier, 1974) and E.C.F. Bird, Shoreline Changes during the Past Century (International Geographical Union, Melbourne, 1978). A regional study is by S. Jelgersma, Holocene Sea Level Changes in the Netherlands (Van Aelst, 1961). An account of tracing older shorelines was presented by J.L. Hough and H.W. Menard (Eds.), Finding Ancient Shorelines (Society of Economic Palaeantology and Mineralogy, 1955). Tidal studies tend to be very mathematical and are rarely very relevant to geomorphological studies. A typical book is E.P. Clancy, The Tides (Anchor, 1969).

### Marine Processes - The Morphology of the Ocean Floor

This branch of knowledge has expanded considerably since the development of the automatic echo-sounder which can provide a continuous record of the depth of the sea. The study has gone hand in hand with the invention of corers which can take samples from the sea-bed and with the development of the theory of plate tectonics.

A descriptive account of the individual oceans is given in the various parts of A.E. Nairn, W.R. Kanes and F.G. Stehli, The Ocean Basins and Margins (Plenum, 1973-78). More general overviews are given by C.A.M. King, Introduction to Marine Geology and Geomorphology (Arnold, 1975) and in G.P. Shepard's two books Submarine Geology (Harper and Row, 3rd edn. 1973) and Geological Oceanography - Evolution of Coasts, Continental Margins and Deep Sea Floor (Heinemann, 1978). Also relevant is M. Sears (Ed.), The Quaternary History of the Ocean Basins (Pergamon, 1967).

Marine Processes - Marine Sediments and Sedimentation

This topic is essentially part of sedimentary petrology, but inasmuch as the inshore deposits are largely derived from the land, they are of interest to geomorphologists. Amongst the more recent books that may be cited were, R.A. Davis, Jr. (Ed.), Coastal Sedimentary Environments (Springer Verlag, 1978) and D.J.P. Swift and H.D. Palmer (Eds.), Coastal Sedimentation (Benchmark Paper in Geology, No.42, Dowden, Hutchinson and Ross, 1978).

An interesting collection of papers resulted from a joint meeting of coastal geomorphologists and marine sedimentologists and edited by J.R. Hails and A.P. Carr, Nearshore Sediment Dynamics and Sedimentation (Wiley, 1975). The deposits found in tidal inlets have attracted considerable attention, partially because of their effect upon channels used for navigation, three recent collections of papers were R.N. Ginsberg (Ed.), Tidal Deposits; a case-book of recent examples and fossil counterparts (Springer Verlag, New York, 1975); G.de V. Klein (Ed.), Holocene Tidal Sedimentation (Benchmark Paper in Geology, No.30, Dowden, Hutchinson and Ross, 1976) and R.A. Davis, Jr. and R.L. Ethington (Eds.), Beach and Nearshore Sedimentation (Soc.Ec.Pal.and Mineral., Spec.Pub.No. 24, 1976).

Marine Processes - Estuaries

Estuaries are intermediate in morphology, sedimentation and marine chemistry between freshwater and the open ocean, gradually changing from the former to the latter. In consequence they show considerable variations in tidal and wave effects as well as in their biology, and have attracted considerable attention from scientists across a wide spectrum of disciplines. There have been several books published on estuaries lately, though most are edited publications. One of the earlier collections of papers was C.H. Lauff (Ed.), Estuaries (Am. Assoc. Arts and Sci. Washington, 1967) which includes material of direct geomorphological interest, together with purely biological and environmental studies. A similar more recent work was V.S. Kennedy (Ed.), Estuarine Perspectives (Academic Press, 1980). Of more specialised physical interest were A.T. Ippen, Estuary and Coastline Hydrodynamics (McGraw-Hill, 1966); M. Wiley (Ed.), Estuarine Processes, 2 Vols,

(Academic Press, 1976) and M. Wiley (Ed.), Estuarine Interactions (Academic Press, 1978). A British example of a multi-disciplinary approach was M.B. Collins, et al., Industrialised Embayments and their Environmental Problems (Pergamon, 1980) which covered various aspects of the environment of Swansea Bay. On the purely morphological aspects of an estuary might be quoted H.R. Wilkins, C. De Boer and A. Thunder, A Cartographic Analysis of the Changing Bed of the Humber, 2 Vols. (Univ. of Hull, Dept. of Geog., Misc. Ser. No.14).

## Marine Processes - Waves and Currents

There is now general agreement on the dominance of waves as coastal agents of erosion and, in part, of deposition, with their range being extended by tidal variations. The role of currents is correspondingly diminished, though their role in sediment transport cannot be ignored. Some treatment of the various kinds of marine waves is included in all standard works of oceanography, especially of physical oceanography. There has also been considerable refinement in the instruments and techniques used to study waves, of which air photography is an important method. Amongst the books devoted to them and their relationship to beaches is W.N. Bascom, Waves and Beaches (Doubleday, New York, 1964) and R.E. Meyer (Ed.), Waves on Beaches and Resulting Sediment Transport (Academic Press, 1972). A special type of wave is the tsunami, often misguidedly called a tidal wave. Such waves are highly destructive and are generated by earthquakes. One book entirely devoted to this was T.S. Murty, Seismic Sea Waves, Tsunamis (Bull. 198, Fisheries Res. Bd.of Canada, Ottawa, 1977).

Currents have not been studied to the same extent, but one such is L.M. Fomin, The Dynamic Method in Oceanography (Elsevier, 1964), which, despite its general-sounding title, was chiefly concerned with currents and the charting of them. On the other hand H. Stommel's The Gulf Stream - A Physical and Dynamic Description (Univ. of California Press, Berkeley and C.U.P., 1958) dealt with one specific current.

## Marine Processes - Constructional Forms (beaches, spits, etc)

These features figure prominently in most texts on coastal geomorphology, but there are several books devoted solely to beaches and allied forms and the processes that form them. Perhaps one of the most comprehensive was P.D. Komar, Beach Processes and Sedimentation (Prentice-Hall, 1976). The work of R.A. Davis and R.L. Ethington (Eds.) (1976) has already been mentioned above, and older papers were collected by J.S. Fisher and R. Dolan (Eds.), Beach Processes and Coastal Hydrodynamics (Benchmark Paper in Geology, No.39, Dowden, Hutchinson and Ross, 1977).

Of the offshore phenomena such as spits, bars, barrier beaches and islands, etc, these are covered by two volumes of papers edited by M.L. Schwartz, Spits and Bars (Benchmark Paper in Geology, No.6., 1972) and Barrier Islands (Benchmark Paper in Geology, No.39, Dowden Hutchinson & Ross, 1977). A more recent issue was S.P. Leatherman (Ed.), Barrier Islands from the Gulf of St.Lawrence to the Gulf of Mexico, (Academic Press, 1979).

Inland, especially on low coasts there are accumulations of sands, often derived from beaches, known as dunes, which have received much detailed study from biologists because they form a special habitat. One useful geomorphological study was W.S. Cooper, Coastal Dunes of California (Mem.Geol. Soc.Am. No.104, 1967). A similar British study was R.K. Gresswell, Sandy Shores in South-West Lancashire (Liverpool Univ. Press, 1953).

## Marine Landforms - Regional Accounts

For Great Britain, the two standard works are by J.A. Steers, The Coastline of England and Wales, (C.U.P., 1948) and The Coastline of Scotland (C.U.P., 1973). The former contains a very useful photographic supplement with 115 plates which was also published separately,together with a further 31 plates, as A Picture Book of the Whole Coast of England and Wales (C.U.P.,1948). The Countryside Commission of Scotland also published a series of studies of Scottish beaches written by A.S. Mather, W. Ritchie and others, between 1975-1978. There were several books published on East Anglian coasts, examples included J.A. Steers, (Ed.) Scolt Head Island (Heffer, 2nd edn., 1960) and the anonymously edited, Orford Ness - A Selection of Maps Mainly by John Norden, Presented to James Alfred Steers (Heffer, 1966). From Dorset came two recent studies, A.D. Canning and K.R. Maxsted, Coastal Studies in Purbeck (Purbeck Press, 1979) and D. Brunsden and A. Goudie, Classic Coastal Landforms of Dorset (Geog. Assoc. and BGRG, 1981). There have also been several coastal studies published by Geography Departments. For the Bristol channel one can cite F.J. North, The Evolution of the Bristol Channel (Nat. Museum of Wales, Cardiff, 2nd edn., 1955) and F. Rowbotham, The Severn Bore, (David and Charles, 1964). For the Irish Sea thereis C. Kidson and M.J. Tooley (Eds.) The Quaternary History of the Irish Sea (Seel House Press, Liverpool, now Wiley, 1977).

## Geomorphological Periodicals

Most general geographical and geological journals devote a considerable amount of their space to geomorphological articles and for geography these are listed by C.D. Harris, Annotated World List of Geographical Serials (University of Chicago, 4th edn. 1980).

There are also journals devoted solely to geomorphology and one of the oldest of these is the Zeitschrift für Geomorphologie (Gebruder Borntrager, 1925-40) which was resuscitated in a second series from 1957. The American Journal of Geomorphology (Columbia University, 1938-42) did not survive the war. However in 1976 a new English language journal appeared, Earth Science Processes (Wiley, 1976-80) which became Earth Science Processes and Landforms in 1980. This journal is the official organ of the British Geomorphological Research Group. The Group also issues Technical Bulletins and occasionally Current Research in Geomorphology. Germany and France both have their own journals also; Catena (Catena Verlager, 1973-) and Revue de Géologie Dynamique et Géographie Physique (Masson 1928-39 and 1954-). Since 1979 the publication has reversed the two halves of its title. Also Poland produces Studia Geomorphologica Carpatho-Baltanica (Polish Academy of Science, 1967-).

There has been of late a growing interest in applying geomorphology in a practical way, chiefly in the fields of geology and geography. Applied geology is covered by Reviews in Engineering Geology (Geological Society of America, 1962-), Engineering Geology (Elsevier, 1965-) and the Quarterly Journal of Engineering Geology (Geological Society of London, 1967-). There have been two journals called Applied Geography, one appeared in Japan between 1960 and 1964 and the other appeared in 1981 from Butterworths.

## Reference Books on Geomorphology

We will now turn our attention to general reference books on geomorphological materials. Often such books are separated from geomorphological texts in libraries by being placed in reference presses where they are often ignored. Probably the most important reference work for geomorphologists is R.W. Fairbridge (Ed.), The Encyclopaedia of Geomorphology (Reinhold, 1968). It is a massive, comprehensive and scholarly work, whose articles are frequently cited in the bibliographies of articles in the periodical literature. A much smaller, and more specialised book, which is really a dictionary or glossary rather than an encyclopaedia is E. Vollmer (Ed.), Encyclopaedia of Hydraulics, Soil and Foundation Engineering (Elsevier, 1967). The author is unaware of any dictionaries of geomorphology, however the subject is well covered in general dictionaries of geography and geology. Amongst the former are W.G. More, A Dictionary of Geography (Penguin, 6th edn., 1981) and Sir Dudley Stamp and A.V. Clark (Eds.), Glossary of Geographical Terms (Longmans, 3rd edn. 1978). This last volume provides the original usage where it can be traced and subsequent usage, together with a recommendation where ambiguities have crept in. In the geological field may be cited D.G.A. Whitten and J.V. Brookes, The Penguin Dictionary of Geology (Penguin, 1972).

At all stages of geomorphological enquiry there arises the problem of discovering what has already been published on your topic, but for Britain this has been eased by being able to consult K.M. Clayton (Ed.) Bibliography of British Geomorphology (Philip, 1964). For subsequent years one can check entries in Geomorphological Abstracts (1960-65) and its successor, Geo Abstracts. Consolidated indices are published every five years by computer.

Despite the quantitative revolution many geomorphologists remain resolutely qualitative and others consider that geomorphology ends at the base of the regolith. However as geomorphologists enter the applied field and have to talk to engineers and geologists, it is increasingly important that they should be aware of the materials from which landforms are carved or produced by deposition. One of the few books in this field is W.H. Whalley Properties of Materials and Geomorphological Explanation (O.U.P., 1976). Geological geomorphologists have been long aware of the need for a knowledge of petrology and mineralogy. No geomorphologist can seriously afford to ignore these areas, and works such as B. Mason Principles of Geochemistry (Wiley, 1968) and W.G. Ernst Earth Materials (Prentice-Hall, 1969) should not be unknown to him or her.

Chapter Three

# HISTORICAL GEOGRAPHY

J.A. Yelling

Such is the wide scope of this topic that no account can be both comprehensive and enlightening, and compromise is necessary. I have chosen to attack the subject from several directions. First, there is a general discussion of the aims, content and method of historical geography. In the past such an account might have served as the basis for the whole chapter, but in recent decades methodological writings have tended to be less related to the broad mass of substantive work than they were previously. A second approach is through general works and collections of works relating to specific countries or parts of the world. The rest of the chapter is composed around three selected themes with which historical geographers have been much concerned. There is also a brief section devoted to periodicals and bibliographies.

## Scope and Purpose

Historical geography has two main general aims. One of these is to provide an understanding of the geography of the present, the other is to pursue the geographical study of the past. In theory, these aims are complementary, since the best account of the past for geographers of the present must surely be one which has been thoroughly shaped by the application of geographical methods. In practice, however, historical geographers have often had great difficulty in deciding how they should study the past, or contribute to the present, and such methodological considerations have greatly shaped the character of their output. This is made clear in A.R.H. Baker (Ed.) Progress in Historical Geography (David and Charles, 1972), the most important recent review of the state of the field in each of the main practicing countries.

In its short history, historical geography has tended to become wider in scope, whilst individual works have become more specialised. After early studies concerned with the effects of the environment on the development of civilisations, attention was concentrated from the 1930s to the 1950s on the twin concepts of cultural landscape and region, not always clearly separated. Good accounts of the methodological discussions of this period are contained in the chapters on historical geography by H.C. Darby in H.P.R. Finberg (Ed.) Approaches to History (Routledge, 1962) and C.T. Smith in R.J. Chorley and P. Haggett (Eds.) Frontiers in Geographical Teaching (Methuen, 1965).

Fashioned by human action the cultural landscape complements the natural landscape studied by physical geographers. As an object of study it constitutes one of the most tangible expressions of the past in the present, and there is an inevitable emphasis on its historical development. Another important advantage is the applicability of field work, cartographic analysis and air photography, all recognised areas of geographical expertise. Particular attention was paid in Germany to the development of rigorous methods of analysing these types of evidence. The resulting morphogenetical approach was largely responsible for the flowering of German agrarian studies to be discussed in a later section.

In the Anglo-Saxon world landscape studies mostly took a different form with less emphasis on method and classification. H.C. Darby drew attention to various themes which might be followed in the context of the changing English landscape, including the clearance of the woodlands and the drainage of the fens. An even greater influence was exercised by the work of the historian W.G. Hoskins on The Making of the English Landscape (Hodder and Stoughton, 1955), which was extended to form a series of volumes on the counties of England by various authors. Several geographers have contributed to this series. R. Millman has discussed The Making of the Scottish Landscape (Batsford, 1975), whilst outside Britain M.W. Williams The Making of the South Australian Landscape (Academic Press, 1974) is a good example drawing inspiration from both Darby and Hoskins.

In the United States the concept of the cultural landscape was fostered by C.O. Sauer whose work forms the subject of J. Leighly (Ed.) Land and Life (University of California Press, 1963). Sauer particularly emphasised the transformation of the natural landscape by man, and developed research interests in close association with anthropological and ecological studies. The contributions to W.L. Thomas (Ed.) Man's Role in Changing the Face of the Earth (University of Chicago Press, 1955) give some indication of the broader themes within this tradition. Sauer was also associated with the notion of changing cultural landscapes in relation to settlement by different ethnic groups. P.O. Wacher Land and People: A Cultural Geography of Pre-Industrial New Jersey (Rutgers University Press, 1975) is an example of recent work explicitly modelled on this approach.

It will be evident that landscape studies are still very much part of the repertoire of historical geographers, and they have been extended in various ways. Most now take greater account of landscape perception and tastes. Practical questions concerning the recognition and planning of historic landscapes are raised in R.M. Newcomb Planning the Past: Historical Landscape Resources and Recreation (Dawson, 1979). D.W. Meinig (Ed.) in The Interpretation of Ordinary Landscapes (O.U.P., 1979) stresses that one of the aims is to enable a greater enjoyment of landscapes by both layman and professional.

The second theme of the region and 'regional geography' had the advantage of not being confined to the study of any particular set of features. Nearly all historical geography is set within regions of greater or lesser size, but regional geography was conceived as a synthesis of all the features that gave character to area. It soon became clear that this character or 'personality' was shaped not only by physical geography but also, and perhaps more importantly, by patterns of historical development. Increasingly the historical part received lengthier treatment and the physical less. This met with some resistance from those who considered that it weakened the methodological unity of geography. Such a view was taken by R. Hartshorne in The Nature of Geography (Association of American Geographers, 1939). He considered all narrative treatment to be history, but left room for one type of historical geography, namely the study of past periods as though they were a geographical cross-section through time. This approach became particularly important in Britain, where H.C. Darby's Domesday Geography of Eastern England (C.U.P., 1952) came to be considered the classic work of the type. With later volumes on other regions, it reconstructed the geography of England c.1086 from Domesday Book. The findings were later brought together in Domesday England (C.U.P., 1977). Another influential reconstruction of the past geography of a particular period was R.H. Brown Mirror for Americans: Likeness of the Eastern Seaboard 1810 (American Geographical Society, 1943). In so far as historical geographers wished to contribute to an understanding of the present they could do so through a sequence of cross-sectional views, perhaps linked by explanatory narratives.

The emphasis on regional geography is clearly shown in the two text books which formed the basis of courses through which most current historical geographers in the Anglo-Saxon world were trained. These were H.C. Darby (Ed.) An Historical Geography of England before 1800 (C.U.P., 1936) and R.H. Brown Historical Geography of the United States (Harcourt Brace, 1948). The method of cross-sections and connecting narrative was still used by Darby in organising A New Historical Geography of England (C.U.P., 1973). Thus although the regional approach is now much less prominent it still occupies a place. This is also illustrated by new works, like D.W. Meinig The Great Columbia Plain (University of Washington Press, 1968) which explicitly deal with patterns created by the interplay of many factors rather than with individual processes.

In many ways the 1950's and 1960's saw a new beginning in historical geography as in other parts of the subject. This took both a hard and a soft form, of which the soft came earlier and had greater effect. It involved a shift away from concern with region and landscape to a more direct attack on economic and social organisation, institutions and processes. Historical geographers became more specialised in agriculture, industry or urban matters, particularly in Britain. Studies produced along these lines in the 1960's formed the basis of R. Dodgshon and R.A. Butlin (Eds.) An Historical Geography of England and Wales (Academic Press, 1978). In France and the United States the regional interest was preserved perhaps to a greater degree, but even so much more topical specialisation was also apparent. At the same time cross-sections and like methods became less important.

New beginnings in their hard form centred on attempts to transform geography into a discipline more akin to economics. The core of the subject was to be the formulation of general spatial principles largely obtained and expressed through abstract models. Historical geography was to be simply the application of those principles to the past. This methodology with its emphasis on theory and quantitative methods has been termed the 'new geography', and is equivalent in many ways to the 'new economic history' which developed at a similar time. This terminology is, however, dangerous if it is taken to mean that the only alternative to the 'new geography' is the study of landscape and regional geography.

The 'new geography' did have applications in historical geography, and several examples are given in A.R.H. Baker (Ed.) Progress in Historical Geography (David and Charles, 1972), which was published when this movement was at its height. Others will be mentioned later in this chapter. In many respects the advent of the 'new geography' represented a defeat for the ideas which historical geographers had sought to advance. The quest for spatial principles of a highly generalised form drew attention away from the complex chains of development which historical geographers used to construct their view of the world. There seemed much less point in looking to the past as an aid to understanding the present.

During the last decade, therefore, historical geographers have seen their more specialised output win greater acceptance from historians, but they have had to fight hard to re-establish their position in geography. They began by joining with others in stressing perception and the 'behavioural' response to situations. This movement did a great deal to prevent the 'new geography' settling into an extreme rigidity and it enabled historical geographers to re-emphasise the importance of cultural traits and values. This is reflected in H.C. Prince's account of 'Real, Imagined and Abstract Worlds of the Past' in Progress in Geography 3 (Arnold, 1971).

The most recent development has been a greater emphasis on social matters which are no longer treated solely as an appendage to economic organisation. Social questions have also been emphasised in the first explicitly marxist approaches in historical geography as in D. Gregory _Ideology, Science and Human Geography_ (Hutchinson, 1978). Above all, historical geographers have been able to point to a large number of new works which have been produced since the 1960's, and which we must now proceed to discuss. Although areas of weakness remain, their availability puts historical geography in a better position to fight ahistorical trends than it was when the 'new geography' began.

_General Survey_

This section is confined to works in English which deal with Europe, North America or Australasia. C.T. Smith _An Historical Geography of Western Europe before 1800_ (Longmans, 1967) remains the best work on this theme. The volumes by N.J.G. Pounds _An Historical Geography of Europe 450 BC - AD 1330_ (C.U.P., 1973) and _1500-1840_ (C.U.P., 1979) are useful works of reference, particularly on urban growth and industries. C. Delano-Smith provides a new treatment of _Western Mediterranean Europe_ (Academic Press, 1979) and F.W. Carter _An Historical Geography of the Balkans_ (Academic Press, 1977).

For individual European countries, H.D. Clout _Themes in the Historical Geography of France_ (Academic Press, 1977) provides a valuable survey of most aspects of French historical geography. M. Jones _Finland:Daughter of the Sea_ (Dawson, 1977) has a rather more specialised theme, as does D. Stanislawski _The Individuality of Portugal_ (University of Texas, 1959). Works in the landscape tradition include A.M. Lambert _The Making of the Dutch Landscape_ (Academic Press, 1971) and F.H.A. Aalen _Man and the Landscape in Ireland_ (Academic Press, 1978).

British historical geography is now best covered by H.C. Darby (Ed.) _A New Historical Geography of England_ (C.U.P., 1973) and R. Dodgshon and R.A. Butlin (Eds.) _An Historical Geography of England and Wales_ (Academic Press, 1978). A.R.H. Baker and J.B. Harley _Man Made the Land_ (David and Charles, 1973) offers a simpler, visually illustrated, introduction. A.R.H. Baker, J.D. Hamshere and J. Langton (Eds.) _Geographical Interpretations of Historical Sources_ (David and Charles, 1970) reproduces some of the best periodical articles then available. J. Patten _Pre-Industrial England_ (Dawson, 1979) does the same for the seventeenth century, with an introduction. E. Pawson has provided a general historical geography of eighteenth century Britain in _The Early Industrial Revolution_ (Batsford, 1979). A companion volume is P.J. Perry _A Geography of Nineteenth Century Britain_ (Batsford, 1975).

R.H. Brown's _Historical Geography of the United States_ (Harcourt Brace, 1948) remains the most comprehensive work on this subject, especially valuable for its regional studies. Recent volumes, however, now provide a new general foundation for the study of historical geography in the United States. These are D. Ward (Ed.) Geographic Perspectives on America's Past (O.U.P., 1979), J.R. Gibson (Ed.) European Settlement and Development in North America (Dawson, 1978), and R.E. Ehrenberg (Ed.) Pattern and Process: Research in Historical Geography (Howard U.P. 1975). All three works consist of separate essays, many of which have either appeared in periodicals or are derived from larger studies. Together, they constitute a representative view of the activities of United States historical geographers, and the Gibson volume also extends to Canada. Inevitably, the major theme is that of pioneer settlement and rural development, but there is some mention of the urban/industrial world. The Ehrenberg volume also contains an introduction by the editor to material available in the United States National Archives.

R.C. Harris and J. Warkentin Canada Before Confederation (O.U.P., 1974) provides an introduction to that period, whilst A.H. Clark and D.Q. Innis have contributed an historical section to J. Warkentin (Ed.) Canada: A Geographical Interpretation (Methuen, 1968). In similar fashion, historical geographers contributed to G.H. Dury and M.I. Logan Studies in Australian Geography (Heinemann, 1968). For Australia, however, pride of place must now go to J.M. Powell and M. Williams (Eds.) Australian Space, Australian Time (O.U.P., 1975), a collection of essays which provides a new benchmark for the study of historical geography in that country.

European Agrarian Studies

Very few of the more specialised agrarian studies have yet transcended national boundaries. There does exist, however, a Permanent European Conference for the Study of the Rural Landscape and published volumes of papers from its biennial meetings give an indication of the type of work carried out by geographers in the various countries. Two good examples are the proceedings reported in 'Morphogenesis of the Agrarian Cultural Landscape' Geografiska Annaler 43 (Stockholm, 1961), and R.H. Buchanan, R.A. Butlin and D. McCourt (Eds.) Fields, Farms and Settlements in Europe (Ulster Folk and Transport Museum, 1976). A.R.H. Baker (Ed.) Progress in Historical Geography (David and Charles, 1972) includes accounts of work in Britain, France, Germany, Scandinavia and the U.S.S.R., and is perhaps the best single starting point for more detailed enquiries. M.L. Parry Climatic Change, Agriculture and Settlement (Dawson, 1978) is a work of wide geographical reference on a single theme.

Historical geographers mostly began work too late to contribute to the first major syntheses of European agrarian structures. An important exception, however, was R. Dion's Essai sur la Formation du Paysage Rural Francais (Arrault,

1934), a work still worth reading in its original. French scholars did not follow this lead, but concentrated instead on regional studies. Within this format much attention has been devoted to the development of agrarian structures and land use since the eighteenth century, and to the question of rural depopulation. An important volume by E. Juillard, A. Meynier, X.de Planhol and G. Sautter Structures Agraires et Paysages Rurales (Université de Nancy, 1957) reviews work in these fields. Later references are best obtained from H. Clout (Ed.) Themes in the Historical Geography of France (Academic Press, 1977). Two important recent works on France by English scholars are H. Clout Agriculture in France on the Eve of the Railway Age (Croom Helm, 1980) and G. Meirion-Jones La Maison Traditionelle (Paris,C.N.R.S., 1978) a bibliography of vernacular architecture. R. Dion Histoire de la Vigne et du Vin en France des Origines au XIX$^{e}$ Siècle (Flammarion, 1959) should be noted. It may be compared with H. Hahn Die Deutschen Weinbaugebiete (Bonn Geogr. Inst.,1956) and D. Stanislawski Landscapes of Bacchus: The Vine in Portugal (University of Texas Press, 1970).

Mention has already been made of the significance of German work in this field. A. Mayhew Rural Settlement and Farming in Germany (Batsford, 1973) provides an excellent introduction in English. G. Schwarz Allgemeine Siedlungsgeographie (De Gruyter, 1966) is still a useful reference. M. Born Die Entwicklung der Deutschen Agrarlandschaft (Wissenschaftliche Buchgesellschaft, Darmstadt, 1974) is the best short survey of agricultural development, whilst H-J Nitz Historisch-genetische Siedlungsforschung (same date and publisher) is a collection of many of the most important papers. Individual monographs cannot be dealt with here, but mention may be made of I. Leister Peasant Openfield Farming and its Territorial Organisation in County Tipperary (Marburger Geogr. Schr., 1976) as an interesting example of German methods applied to Ireland. These methods have also been adopted and extended by Scandinavian scholars, who have particularly introduced detailed measurement or metrology in studies of both field systems and settlement. This work is best approached through the chapter by Helmfrid in Baker (1972) cited above.

In Britain much work on field systems and related issues was brought together in A.R.H. Baker and R.A. Butlin (Eds.) Studies in Field Systems in the British Isles (C.U.P., 1973). This consists mainly of detailed regional surveys, but general formative factors are discussed in a conclusion. An excellent review of those factors is provided by R.A. Dodgshon The Origin of British Field Systems (Academic Press, 1980), which also contains an entirely new interpretation of the infield-outfield system. A comprehensive study of rural settlement is now available in B.K. Roberts Rural Settlement in Britain (Dawson, 1977), whilst J.A. Yelling gives a general view of Common Field and Enclosure in England 1450-1850 (Macmillan, 1977).

A. Harris The Rural Landscape of the East Riding of Yorkshire 1700-1850 (University of Hull, 1961) is a model study of changes in agricultural structure and land use at the regional level. The enclosure and drainage of fenlands has been the subject of two notable studies by H.C. Darby The Draining of the Fens (C.U.P., 1940) and M. Williams The Draining of the Somerset Levels (C.U.P., 1966). R.A. Donkin brings together his work on the Cistercians in The Cistercians: Studies in the Geography of Medieval England and Wales (Inst. of Medieval Studies Toronto, 1978). C. Taylor has examined Fields in the English Landscape (Dent, 1975) and P. Smith Houses in the Welsh Countryside (H.M.S.O., 1975). I. White has produced a fine study of Agriculture and Society in Seventeenth Century Scotland (John Donald, 1979).

Most of the works so far mentioned refer to changes in economic production and social organisation as well as agrarian structure. D.B. Grigg The Agricultural Revolution in South Lincolnshire (C.U.P., 1966) is notable for its emphasis on rent and productivity. D. Thomas Agriculture in Wales during the Napoleonic Wars (University of Wales, 1963) is concerned mainly with production as is P.J. Perry British Farming in the Great Depression 1870-1914 (David and Charles, 1974). D. Mills English Rural Communities (Macmillan, 1973) is a collection of papers on open and closed villages and other aspects of settlement and society. Finally, many threads in the historical geography of rural Britain are apparent in the papers contained in H.S.A. Fox and R.A. Butlin (Eds.) Change in the Countryside: Essays on Rural England 1500-1900 (Institute of British Geographers, 1979) and M.L. Parry and T.R. Slater (Eds.) The Making of the Scottish Countryside (Croom Helm, 1980).

## Colonisation and Frontier Settlement

Work on these themes has formed the major part of the output of historical geographers in N. America and Australasia, and it is best introduced through the books already mentioned in the general survey. With them may be cited J.M. Powell Mirrors of the New World (Dawson, 1977) which discusses both Australia and N. America concentrating on images and image makers in the settlement process. Most studies are, however, more limited in scale, and there is a considerable indentity of themes, although topical specialisation has tended to increase. The approach adopted here is to begin by picking out some main lines of inquiry, illustrated by a few works, leaving other publications to be summarised at the end. Australasia will be treated separately.

European exploration and discovery are natural geographical themes, very often linked in modern studies with questions of the images and reality of the new lands, and the nature of indigenous society. These matters are discussed in several works by C.O. Sauer including The Early Spanish Main

(C.U.P., 1966) and Sixteenth Century North America (University of California Press, 1971). J. Warkentin includes such topics in The Western Interior of Canada: A Record of Geographical Discovery 1612-1917 (McClelland and Stewart, 1964). Other examples are D.R. McManis European Impressions of the New England Coast 1497-1620 (Chicago Research Papers, 1972) and J.L. Allen Passage through the Garden: Lewis and Clark and the Image of the American North West (University of Illinois Press, 1975)

Considerable attention has always been paid to the role of various immigrant groups in shaping landscape, economy and society. T.G. Jordan thus examines German Seed in Texas Soil (University of Texas Press, 1966) and R.L. Gerlach Immigrants in the Ozarks (University of Missouri Press, 1976). J.J. Mannion provides a good example of this theme in Irish Settlement in Eastern Canada (University of Toronto, 1974). J.G. Rice Patterns of Ethnicity in a Minnesota County 1880-1905 (University of Umea, 1973) is concerned with Swedish settlement, whilst J.R. Gibson has tackled Imperial Russia in Frontier America (O.U.P., 1976). A related theme is that of the different economies which often succeeded each other with increasing intensification of land use. Thus D.J. Wishart has studied The Fur Trade of the American West 1807-1840 (Croom Helm, 1979), and A.J. Ray Indians in the Fur Trade (University of Toronto Press, 1974) deals with the lands south west of Hudsons Bay 1660-1870. At the other extreme M.P. Conzen discusses Frontier Farming in an Urban Shadow (University of Wisconsin, 1971). Landholding, and the impact on the landscape of methods of land partitioning and survey are discussed in works like R.C. Harris The Seigneurial System in Early Canada (University of Wisconsin Press, 1966), W.D. Pattison The Beginnings of the American Rectangular Land Survey System 1784-1800 (University of Chicago, 1957) and N.J.W. Thrower Original Land Subdivision: A Comparative Study of the Form and Effect of Contrasting Cadastral Surveys (Association of American Geographers, 1966). Folk cultures and landscapes, a theme developed by F. Kniffen, are discussed in H.J. Walker and W.G. Haag (Eds.) Man and Cultural Heritage (Louisiana State University, 1974) and H. Glassie Pattern in the Material Folk Culture of the Eastern United States (University of Pennsylvania, 1978).

Many important regional studies develop some or all of these themes, and others, but it is impossible to deal with their individuality. The work of D.W. Meinig on The Great Columbia Plain (University of Washington Press, 1968) has already been mentioned, and the same author has studied Imperial Texas (University of Texas Press, 1969) and the Southwest (O.U.P., 1971). For the east, there are important studies by D.R. McManis on Colonial New England (O.U.P., 1975) and H.R. Merrens on Colonial North Carolina in the Eighteenth Century (University of North Carolina Press, 1964). J.T. Lemon has produced Best Poor Mans Country: A Geographical Study of S.E. Pennsylvania (Johns Hopkins University Press, 1972) and P.O. Wacher Land and People: A Cultural Geography of Pre-Industrial New Jersey (Rutgers University Press, 1975). The same author has tackled the Muscenetcong Valley of New Jersey (Rutgers University Press, 1968), R.D. Mitchell the Shenandoah in Commercialism and Frontier (University of Virginia, 1971) and J.A. Jackle Images of the Ohio Valley 1740-1860 (O.U.P., 1977). C.V. Earle The Evolution of a Tidewater Settlement System: All Hallow's Parish, Maryland 1650-1783 (University of Chicago, 1975) is a good example of detailed local research. For Canada A.H. Clark produced regional surveys of early Nova Scotia in Acadia (University of Wisconsin Press, 1968) and of Prince Edward Island in Three Centuries and the Island (University of Toronto Press, 1959). C.G. Head has examined Eighteenth Century Newfoundland (McClelland and Stewart, 1976) whilst E. Ross Beyond the River and the Bay (University of Toronto Press, 1970) deals with Manitoba in 1811. J.D. Wood has edited a collection of essays by historical geographers in Perspectives on Landscape and Settlement in Nineteenth Century Ontario (McClelland and Stewart, 1975).

Work in Australasia has been naturally much influenced by that in North America, sometimes through the direct impact of individual scholars like A.H. Clark on The Invasion of S. Island New Zealand by People, Plants and Animals (Greenwood, 1949) and D.W. Meinig On the Margins of the Good Earth: The S. Australian Wheat Frontier 1788-1929 (Association of American Geographers, 1963). In recent decades, however, Australian work has been very vigorous, providing some of the best examples of frontier settlement studies. Two of the pioneers who helped to make this possible are celebrated in collections of articles by J. Andrews (Ed.) Frontiers and Men: A Volume in Memory of Griffith Taylor 1880-1963 (Melbourne University Press, 1966) and F. Gale and G.H. Lawton (Eds.) Settlement and Frontier: Geographical Studies presented to Sir Grenfell Price (O.U.P., 1969).

Questions of land appraisal, images and reality, and the contrasts between ideology and reality in official and private settlement schemes have been dominant issues. They are considered in R.L. Heathcote Back of Bourke (Melbourne University Press, 1965) and J.M. Powell The Public Lands of Australia Felix (O.U.P., 1970). They are prominent in the papers contributed to J.M. Powell (Ed.) The Making of Rural Australia (Sorrett Publishing, 1974). The frontier theme is taken up in M.E. Robinson The New South Wales Wheat Frontier 1851-1911 (Australian National University, 1976) and T.M. Perry Australia's First Frontier: The Settlement of New South Wales 1788-1829 (Melbourne University Press, 1963). Other regions are studied in M. Williams The Making of the South Australia Landscape (Academic Press, 1974) and D.N. Jeans An Historical Geography of New South Wales to 1901 (Reed Education, 1972).

Towns and Industry

Historical geography does not yet reflect the interest in urban and industrial matters that characterises modern human geography. Indeed, much of the historical work in these fields has come from those who would not consider themselves primarily historical geographers. In the last decade some attempt has been made to remedy this situation, especially in urban historical geography, but obvious gaps remain. It is essential that these be filled if historical geography is to finally reestablish its proper role vis à vis the rest of human geography.

Early work on towns was much concerned with plan analysis, and in the anglo-saxon world was promoted by scholars with close links with Germany. R.E. Dickenson The West European City (Routledge, 1951) and M.R.G. Conzen Alnwick, Northumberland (Institute of British Geographers, 1960) are examples at different scales. More recently, C.F. Forster Court Housing in Kingston upon Hull (University of Hull, 1972) is a good example of detailed work on nineteenth century housing using cartographic and visual evidence. British Townscapes have been studied by E. Johns (Arnold, 1965) and by M. Aston and J. Bond in The Landscape of Towns (Dent, 1976).

Important contributions to the understanding of pre-industrial towns are contained in P. Wheatley The Pivot of the Four Quarters (Beresford, 1971) on the early Chinese city, and P. Wheatley and T. See From Court to Capital (University of Chicago Press, 1978) on Japanese urban origins. J.E. Vance has developed several wide-ranging theories, one of them in The Merchants World (Prentice-Hall, 1970). J. Patten English Towns 1500-1700 (Dawson, 1978) is a thorough study of most non-morphological aspects. R.A. Butlin has papers by himself and others in The Development of the Irish Town (Croom Helm, 1973), whilst I. Adams has investigated The Making of Urban Scotland (Croom Helm, 1978) and H. Carter.The Towns of Wales (University of Wales Press, 1965). P. Schoeller Die deutschen Städte (Franz Steiner, Wiesbaden, 1967) offers a key to German studies in this field. H.F. Carter Dubrovnik: A Classic City State (Academic Press, 1972) should also be mentioned here. Urban geographers turned increasingly in the 1960's to the study of urban systems, and the internal spatial organisation of towns, and it was in these two fields that the methodology and techniques of the 'new geography' were most apparent. Examples of the first type are B. Robson Urban Growth: An Approach (C.U.P., 1973) which deals with nineteenth century Britain, and two works by A.R. Pred The Spatial Dynamics of U.S. Urban-Industrial Growth 1800-1914 (M.I.T. Press, 1966), and Urban Growth and the Circulation of Information: United States System of Cities 1790-1840 (Harvard University Press, 1973). Influential studies of internal spatial organisation are B. Robson Urban Analysis (O.U.P.,1969) which uses Sunderland as a case example, P. Goheen Victorian Toronto 1850-1900

(University of Chicago, 1970), a notable early example of the application of factor analysis, and R.M. Pritchard Housing and the Spatial Structure of the City (C.U.P., 1976) which discusses mobility and the housing market in nineteenth century Leicester.

The best indicators of recent trends in urban studies in Britain are the papers collected in J.W.R. Whietehand and J. Patten (Eds.) Change in the Town (Institute of British Geographers, 1977) and by R.J. Dennis in The Victorian City (Institute of British Geographers, 1979). For Australia a similar role is performed by J.M. Powell (Ed.) Urban and Industrial Australia (Sorrett Publishing, 1974), whilst D. Ward Cities and Immigrants (O.U.P., 1971) remains the best introduction to urban historical geography in the United States. Two recent studies of particular types of town are J.D. Porteous on British Canal Ports (Academic Press, 1977) and D.B. Knight on the choice of A Capital for Canada (University of Chicago, 1977). The census material from which many British nineteenth century studies are culled is examined in R. Lawton (Ed.) The Census and Social Structure (Cass, 1978).

Two very welcome books by historical geographers which hopefully form part of a revival of work on industry and transport are E. Pawson Transport and Economy: The Turnpike Roads of Eighteenth Century Britain (Academic Press, 1977) and J. Langton Geographical Change in the Industrial Revolution: Coalmining in South-West Lancashire 1590-1799 (C.U.P., 1979). P. Hall The Industries of London since 1861 (Hutchinson, 1962) was a notable early study. Among industrial geographers K. Warren gives much weight to industrial development in The British Iron and Steel Sheet Industry since 1840 (Bell, 1970), Chemical Foundations: The Alkali Industry in Britain to 1926 (Clarendon Press, 1980), and in his introduction to The Geography of British Heavy Industry since 1800 (O.U.P., 1976). B. Fullerton has reviewed The Development of British Transport Networks (O.U.P., 1975) and P. Lewis developed A Numerical Approach to the Location of Industry exemplified by the Distribution of the Paper Making Industry in England and Wales from 1860 to 1965 (University of Hull, 1969).

G.J.R. Linge Industrial Awakening: A Geography of Manufacturing in Australia 1788-1890 (Nelson, forthcoming) will offer a new basis for Australian studies. No such comprehensive volume is available for North America, but there are good individual studies in J.M. Gilmour Spatial Evolution of Manufacturing in S. Ontario 1851-1951 (University of Toronto Press, 1972), R.G. Leblanc The Location of Manufacturing in New England in the Nineteenth Century (Dartmouth College, 1969) and K. Warren The American Steel Industry 1880-1970 (Clarendon Press, 1973).

Periodicals and Bibliography

Articles by historical geographers appear in a wide variety of periodicals, including most of the national geographical series. Special mention should be made of the *Journal of Historical Geography* which has good coverage of the anglo-saxon world and occasional articles on Europe. There is also an *Historical Geography Newsletter* published from the California State University at Northridge. The best sources for book reviews are the *Journal of Historical Geography* and the *Geographical Review*. D.R. McManis has produced a *Bibliography of Historical Geography* for the United States (Eastern Michigan University, 1965), and M.T. Wild a *Register of Research in Historical Geography in Britain* (Geo Abstracts, 1980). *Geo Abstracts Series D* offers the best coverage of historical geography amongst the general bibliographies.

Chapter Four

# AGRICULTURAL GEOGRAPHY

W.B. Morgan

The foundations of modern agricultural geography were laid in the 1950s with the development of the subject as a part of economic geography, concerned in the first instance mainly with production processes and the farm as the decision-making unit, and later examining a variety of aspects including consumption, environment and the role of government. These aspects have been discussed in broad terms in Symons Agricultural geography (1967), Gregor Geography of agriculture: themes in research (1970), Morgan and Munton, Agricultural geography (1971) and Tarrant Agricultural Geography (1974). Earlier work had been mainly concerned with the relationship between farming and environment, but had seen development in classification method and statistical analysis, notably in Baker, Agricultural regions of North America (1926-32), and in studies of spatial relationships and agricultural land use zonation based on the work of Von Thünen Der Isolierte Staat in Beziehung auf Landwirtschaft und Nationalökonomie (1826), the first major attempt to create a theory of economic location. Most of the latter were done by economists, but amongst the geographers we may cite Jonasson, Agricultural regions of Europe (1925). In the 1950s interest in spatial analysis in agricultural geography grew, notably with Garrison and Marble, The spatial structure of agricultural activity (1957), but major summaries of work in this field did not appear until the next decade with Chisholm Rural settlement and land use (1962, 3rd edn.1979), Harvey, Theoretical concepts and the analysis of land use patterns in geography (1966) and later Found, A theoretical approach to rural land-use patterns (1971). A useful summary of theories and model-making in agricultural geography appears in Henshall Models of agricultural activity (1967). Much research in agricultural geography still concentrated on production, but at the farm level, concentrating at first on regional delimitation as in Birch, Observations on the delimitation of farming-type regions, with special reference to the Isle of Man (1954) but eventually developing farm based studies for the purpose of spatial analysis as in Blaikie Spatial organization of agriculture in some north Indian villages (1971) or in a more functional approach using techniques developed by agricultural economists. Coppock in The geography of agriculture (1968) declared "the stimulus to research in agricultural geography has frequently been curiosity about the agricultural characteristics of some territory or about the spatial distribution of some crop or class of livestock. It may be observation of some localised feature... or a general impression of some regional contrast in land use or type of farming (or)... the desire to verify some hypothesis..." But work had already expanded into decision making particularly with the seminal Wolpert The decision process in a spatial context (1964) and included studies not just at the farm but the government level. Agricultural geographers examined a wide range of inputs, husbandry, output, marketing, transport systems and commercial distribution. They have contributed to studies of land tenure, demography, political and social science and economic development. The subject is no longer the rather well defined discipline which it seemed in the 1950s but has spread into and overlapped a very wide range of scientific and social studies.

The most important problems which may be regarded as central to the subject concern attempts to unify its many facets into a single discipline and difficulties in reconciling the various scales at which data are collected and generalizations are made. The subject also shares with the rest of human geography a research base in the real world where data are gathered, hypotheses tested and where studies depend mainly on inductive reasoning. Yet there is also an interest amongst many of its practitioners in constructing theory from deductive methods, initially to predict optimum distributions in totally unreal constructs and more recently to seek closer approximations to reality by satisficer models and techniques involving a behavioural approach. Yet the difficulty remains that even with a study focus firmly within economic geography, the subject is still concerned with cultural and biological relationships. Not all farm-firm managers see their main objective as profit maximisation, but even those who do are concerned essentially with operating a system of living things: crops, livestock, labour, trees, hedge plants, weeds, pests, fungi, bacteria, viruses and the various living things in soil, some of which appear as pests and some of which play important roles in maintaining the nutrient supply and soil structure.

First hand data are preferred to answer many of the important questions raised in agricultural geography. Such data are or should be collected under conditions which relate clearly to the research problem and in themselves represent an original contribution to knowledge.

Most first hand data must be obtained on farms, which are the chief decision making units (decisions affecting agriculture are also importantly made at other levels, notably by governments and by international bodies such as the World Bank). Other sources of first hand data include purchasing organisations, markets, suppliers of inputs, transport firms and agricultural advisers, both private and government. Most agricultural geographers have limited research resources and even with the larger fundings obtainable from some research agencies, making possible the use of field assistants, still find it difficult to conduct their research over any considerable area such as, for example, the whole of Wales. They therefore find it hard to generalise over areas large enough to understand the broader implications of geographical distribution and are rarely, when using first hand data, in a position to advise on national policy. Moreover, they encounter problems of change of scale in asking questions at the farm-firm level which relate to issues seen in a broader areal context, together with problems of composition, i.e. adding the parts in order to view the whole, as noted in Chisholm Problems in the classification and use of farming type regions (1964).

Farm data involve the use of social survey techniques for most economic and social aspects, and physical measurement techniques for biological aspects. Simple observation and area measurement can provide land use data. Important problems in such data collection include the reluctance of most farmers to supply major social survey items such as finance, difficulties of access at busy times of the year, seasonality including some year by year operational overlap, identification of farmer's development objectives and their relationship to research observation, the use of bought-in feeding stuffs, assessment of land quality, allocation of uses of lumpy or fixed capital factors to enterprises, identification of enterprises vis-à-vis crops and livestock in the husbandry system, and even in some cases identification of farms, all their fields and the firm of which they may be only a part. Studies based on data collected in the field have examined a variety of aspects of farming activity in relation to location including farm size (Edwards The effect of changing farm size upon levels of farm fragmentation: a Somerset case study, 1978) and enterprise specialization (Bowler Factors affecting the trend to enterprise specialization in agriculture: a case study in Wales, 1975). In Third World countries physical access difficulties may eliminate large numbers of farms from a study and problems of local survey may restrict interviews to groups at the village rather than the farm level.

Sampling is essential in most first hand data collection and sampling design is extremely important, being recognised more especially by Wood Use of stratified random samples in land use study (1955), Blaut, Microgeographic sampling: a quantitative approach to regional agricultural geography (1959), Birch, A note on the sample-farm survey and its use as a basis for generalized mapping (1960) and Belshaw and Jackson, Type-of-farm areas: the application of sampling methods (1966). In some cases as Clark and Gordon have noted in Sampling for farm studies in geography (1980) neither the census material nor the probability of selection proportional to farm area will suffice and techniques involving clustering and stratification have to be combined.

On the farm area data are often the easiest to obtain together with numbers of livestock at various ages, but area and herd are often not the most satisfactory method of measuring the relative importance of enterprises nor of classifying farms. In consequence agricultural geographers have frequently been driven to converting area data into something which can provide a better index of husbandry or business efficiency. Yields and calorific or joule values may be used (including starch equivalence), but often the yield values available are unreliable. The point is discussed by Coppock (1968) cited above. Standard man days, thought to be independent of variations in management efficiency and yet, as a major input, to reflect better than production indices the effects of decision making in farming systems, are the most commonly preferred index. Their use was developed by Ashton and Cracknell, Agricultural holdings and farm business structure in England and Wales (1960-61) and Napolitan and Brown, A type of farming classification of agricultural holdings in England and Wales according to enterprise patterns (1963). Together with livestock units, a method of obtaining equivalence in numbers between different kinds and ages of livestock introduced by Bennett Jones in The Pattern of farming in the East Midlands (1954), they continue to be used for analytical purposes today. The use of financial data has mostly been left to agricultural economists. There are problems of comparison from year to year due to changes in values and difficulties in the determination of profit and the allocation of inputs and returns to given enterprises. Cruikshank and Armstrong, Soil and agricultural land classification in County Londonderry (1971) have used gross margin analysis in relation to soil properties but the technique has limitations when used for areal comparison.

The seasonality of agricultural activities has often provided problems for research workers in both the collection and the analysis of data. The seasonal programming of farms is subject to regional differences as noted in Curry Regional variation in the seasonal programming of livestock farms in New Zealand (1963). Some farm enterprises have a periodicity which extends from one season to another whilst others occupy differing parts of a given growing season. Timing of visits needs therefore to be done with care as does the allocation of inputs and production data to given activity periods. Over longer periods the cyclic nature of much agricultural production offers problems of comparison where data are only available for given years.

Official data, usually collected in agricultural censuses, are the chief second hand source of information, but in most countries they are rarely available as individual farm returns. Most such data are published by rather large administrative areas or available from government archives or statistics offices for smaller units. Mostly they are collected for purposes very different from those of the research worker in agricultural geography and contain limitations for research use which are sometimes severe. Problems of using such data at the national level have been discussed by Coppock in his An Agricultural Atlas of England and Wales (1964, 2nd edition 1976), a pioneer study using official data and techniques of enterprise-combination analysis, and An Agricultural Atlas of Scotland (1976). Problems in relating the areal basis of such data to farming have been discussed by Coppock in The parish as a geographical-statistical unit (1960) and The cartographic representation of British agricultural statistics (1965), which also examines broader issues.

In great Britain data at the parish level for England and Wales are available from the Ministry of Agriculture, Fisheries and Food (MAFF), Agricultural Census Branch at Guildford and for Scotland from the Department of Agriculture for Scotland (DAS). County returns are published by both the MAFF and the DAS. Useful summary volumes are issued, such as Farm incomes in England and Wales and Farm management survey: gross margins report and there are also map "summaries" in Type of Farming Maps of England and Wales (based on the agricultural census returns for June 1965). Hunt and Clark have provided an excellent introduction to a variety of British and some other data sources in their historical review of The State of British Agriculture 1965-1966 (1966). Farm management data at a regional level based on sample surveys are published by the agricultural faculties of British universities. Common Market data sources include the Bulletin Général de Statistiques, the Statistique Mensuelle du Commerce and the Graphiques et Notes Rapides sur la Conjoncture de la Communauté.

World summary, regional and country-by-country data are published annually (usually two years behind) by the Food and Agriculture Organization of the United Nations in Rome (FAO). Their most important publications include their annual Production Yearbook, Trade Yearbook, the Monthly Bulletin of Agricultural Economics and Statistics, annual review of The State of Food and Agriculture, the FAO Commodity Review and the periodic World Food Survey.

Typology: types of farm, farming areas, land

Agricultural typology is essential for the ordering of data when used as indices of criteria for agricultural qualities. For the most part typology should be related to problems and the techniques needed to solve them, but there have been attempts to create general typologies for the purpose usually of regional comparison at both national and world scales. Such typology, despite criticism that it is of little practical use and may even have inhibited research, has influenced some research workers to standardise their data classes for comparative purposes. Often typology has been restricted to the use of sample farm types rather than a systematic analysis of a population or a sample group. Probably the most successful example of this technique was Dumont, Types of rural economy: studies in world agriculture (1957, originally Economie agricole dans le monde, 1954). The development of a world-wide typology of agriculture has been a major work of the International Geographical Union. An outstanding contribution has come from the work of Kostrowicki in A geographical typology of agriculture: principles and methods (1964), Agricultural typology (1969) and The typology of world agriculture: principles, methods and model types (1972). Polish typological research has been a major contribution to agricultural geography. More recent examples are the work of Falkowski An attempt at the typology and regionalization of agriculture in the Bydgoszcz - Torun agglomeration (1977) and Stola An attempt to apply typological methods in a comparative study of agriculture in Belgium and Poland (1977). Attempts have been made to replace current typological techniques by multivariate analysis as in Tarrant's use of trend surfaces in Some spatial variations in Irish agriculture (1969). A useful summary of earlier work appeared in 1964 with the publication of the IGU Symposium in agricultural geography: the geographical typology of agriculture. Interest in classification is concerned not only with types of farm, land use and enterprises but also land quality where a great deal of pioneer work was done by Stamp The land of Britain: its use and misuse (1948, 3rd edition 1962) and The measurement of land resources (1958). Land classification or capability has been importantly developed in the MAFF, reported by Hilton A new approach to agricultural land classification for planning purposes (1962) and An approach to agricultural land classification in Great Britain (1968). The application of principal components analysis to the problem of land classification has been attempted in Munton and Norris The analysis of farm organisation: an approach to the classification of agricultural land in Britain (1969). In the Soviet Union land potential classification for production has been developed by geographers such as Zvorykin, Scientific principles for an agroproduction classification of lands (1963).

The development of techniques of data analysis and grouping in order not only to classify but to map regional groupings has been a fundamental aspect of the work. An important critique of earlier methods of regional delimitation is Buchanan, Some reflections on agricultural geography (1959). Discussion of problems of regionalisation at the world scale is in Grigg The agricultural regions of the world: review and reflections (1969). Whilst a variety of methods is now available probably the most important step forward was made by Weaver whose method of enterprise combination classification has been the most widely used technique for classification and regional mapping in agricultural geography. Of his several papers on the subject the most fundamental were Crop-combination regions in the Middle West (1954) and Livestock units and combination regions in the Middle West (1956) which was written with others. The technique was developed by Thomas Agriculture in Wales during the Napoleonic Wars (1963) and applied with considerable success by Coppock Crop-livestock and enterprise combinations in England and Wales (1964). All analyses of this kind face problems in evaluating aspects of temporal change varying from farm to farm and region to region. An attempt to develop an objective technique of probability classification applied to temporal changes is in Adams Changing crop combinations and agricultural exports: Upper Midwest response variability (1977). Factor analytic methods have been applied to the problem of crop and livestock combinations by Henshall and King Some structural characteristics of peasant agriculture in Barbados (1966).

Studies of actual types of farming by geographers have been few in the Western world. Amongst the more outstanding have been Simpson The Cheshire grass-drying region (1957) and Erickson The broken Cotton Belt (1948), although even these emphasized the region rather than the forms of agricultural activity. In the Third World the distribution of farming types is a major theme of Manshard Tropical agriculture (1974). Shifting agriculture in particular has been the subject of considerable research, much of it summarised by Watters The nature of shifting cultivation (1960). Most studies of farming activity as such lie outside geography and have been produced by agriculturalists, particularly by agricultural economists.

## Spatial organisation

The organisation of farming in space has received considerable treatment not just in terms of a re-examination of Von Thünen's model of land use, but in terms of a practical examination of urban influences on agriculture, the linkage between farm and market, the evolution of production systems in relation to market and capital source and the arrangement of farming activities around villages, particularly in Third World countries. Gasson The influence of urbanisation on farm ownership and practice (1966) was a major study showing the relation between urbanization, land ownership and enterprise choice, whilst Best and Gasson The changing location of intensive crops (1966) showed the changing geography of horticulture. Sinclair Von Thünen and urban sprawl (1967) discussed the common inversion of Von Thünen's predicted pattern of agricultural intensity around Western cities, although Hall's introduction to Von Thünen's model - Von Thünen's isolated state (1966) - showed that the idea of declining levels of intensity outwards from the city applied only where one kind of production system was involved. The evolution of a spatial system of production has been examined in Harvey Locational change in the Kentish hops industry (1963). Work on farm-market linkages developed in the 1950s, notable contributions coming from Barnes The evolution of salient patterns of milk production and distribution in England and Wales (1958) and Simpson Milk production in England and Wales: collective marketing (1959). Smith examined the effect of the market on changes in land use in Marketing-farm linkages and land use change: a Quebec case study (1974) and has usefully summarised the development of market studies in agricultural geography up to the mid-1970s in The farm-market linkage: a note (1975). Special aspects such as livestock marketing have been treated as in Carlyle, Store stock marketing by small farms in the crofting counties (1978) and Livestock markets in Scotland (1975). Increasingly interest has focussed on the spatial implications of changes in marketing and changes in the supply of inputs. Some aspects of farming, particularly vegetable production, poultry and pigs and certain crops for processing such as sugar beet are affected by contract arrangements and the dominance of the processor or of the feed supplier in the production system. Watts The location of the sugar beet industry in England and Wales 1912-36 (1971) treats the evolution of a cropping and contracting system and White and Watts treat the evolution of an intensive poultry system in The spatial evolution of an industry: the example of broiler production (1977). Contract farming generally is discussed by Hart Geographical aspects of contract farming (1978), Dalton discusses cropping and freezer contracts in Peas for freezing: a recent development in Lincolnshire agriculture, corporate strategy and agribusiness is the subject of Leheron and Warr, Corporate organisation, corporate strategy and agribusiness development in New Zealand (1978), whilst the growing interest of corporate firms in farming, due to the influence of tax structure, has been discussed by Parsons Corporate farming in California (1977). The development of farmers' groups or cooperatives, more especially to take advantage of wholesale methods of purchase or to handle sales more effectively, has also been a subject of study both in the Western World and the Third World. Examples are Bowler Cooperation: a note on governmental promotion of change in agriculture (1972) and Schneider, Brown, Harvey and Riddell Innovation diffusion and development in a Third World setting: the case of the cooperative movement in Sierra Leone (1978).

Marketing of agricultural produce in the Third World has been examined extensively in a considerable literature devoted to a variety of marketing systems including periodic markets. A useful summary for geographers of the early literature is in Bromley Markets in the developing countries: a review (1971). More recent is Smith (ed.) Market place trade - periodic markets, hawkers and traders in Africa, Asia and Latin America (1978). Much less work has been done on the relationship of agricultural production to urban development although a major contribution is Preston Farmers and towns: rural-urban relations in highland Bolivia (1978) and more limited studies have been made elsewhere as Griffin's essay on the zonation of land-use in Testing the von Thünen theory in Uruguay (1973).

A great deal of attention has been devoted to land use zonation around villages and much of this research, developed mainly in the 1950s and 1960s, was summarized in Jackson A vicious circle? - the consequences of von Thünen in tropical Africa (1972).

A final aspect of spatial organisation is the broad grouping of farming activities by regions. Areas specialising in certain forms of production with extremely high levels of concentration are a major feature of the geographical distribution of agriculture. Interest in such regions was an important element in the agricultural geography of the 1920s when Baker developed his method of analysis, but has been revived with a growing interest in their evolution in, for example Spencer and Horvarth How does an agricultural region originate? (1963) and in the realization of the importance of regional planning as in Jensen Regionalization and price zonation in Soviet agricultural planning (1969). In the Third World an important element in the development of agricultural regions has been farmers' migration. Of the several studies of such movement one of the most effective is Hunter Cocoa migration and patterns of land ownership in the Densu Valley near Suhum, Ghana (1963).

## Diffusion of Innovations and Agricultural Change

The geography of agricultural change in broad terms has been discussed mainly at the regional level in studies such as Coppock An agricultural geography of Great Britain (1971) and Morgan Agriculture in the Third World: a spatial analysis (1978). Grigg The agricultural systems of the World: an evolutionary approach (1974) remains one of the most outstanding attempts to examine global change, concentrating particularly on crop diffusion and the evolution of specialized regions of production. A very large part of the interest of geographers in Third World studies has been in agricultural change. Some of this material has appeared in major regional studies such as Thomas and Whittington Environment and land use in Africa (1969) or in reports sponsored by international agencies such as O'Keefe and Wisner, Land use and development (1977). Other research has concentrated on major features of agricultural change such as Chakravarti Green revolution in India (1973), Farmer Green revolution? Technology and change in rice growing areas of Tamil Nadu and Sri Lanka (1977) and Rozlucki The green revolution and the development of traditional agriculture: a case study of India (1977) or has examined the problems arising from change as in Allan and Rosing Disparities in the recognition of indicators of agricultural development - a northwest Indian case (1973). Theories associating agricultural change with population growth have created considerable interest amongst geographers. In South-east Asia one can cite Geertz Agricultural involution: the processes of ecological change in Indonesia (1963), reexamined by Krinks Rural change in Java: an end to involution? (1978).

Boserup's theory of population pressure as an encouragement to agricultural development (The conditions of agricultural growth, 1965) has attracted a wide variety of discussion. Amongst the geographers one can cite Datoo Towards a reformulation of Boserup's theory of agricultural change (1978). A considerable literature examines the relationship of agriculture to population growth in detail, including the problem of carrying capacity. Amongst the more recent examples are Datoo Population density and agricultural systems in the Uluguru Mountains, Morogoro District (1973) and Hervad-Jorgensen Carrying capacity on Tikopia Island (1977).

Change at the farm level in Western countries has attracted the attention of agricultural economists rather than geographers, although a few major studies have been made such as Edwards Complexity and change in farm production systems: a Somerset case study (1980) in which original data for a 15% sample of holdings for 1963 and 1976 are compared. Much more attention has been paid to innovation diffusion, although even here we have to recognise the importance of contributions from non-geographers such as Griliches Hybrid corn: an exploration in the economics of technical change (1957) and Hybrid corn and the economics of innovation (1960). Pioneer work in geography was done by Sauer in the study of crop diffusion in Agricultural origins and dispersals (1952). Broader aspects have been reviewed by Jones, The diffusion of agricultural innovations (1963) and The adoption and diffusion of agricultural practices (1967). Special aspects have been examined by Bowden The diffusion of the decision to irrigate (1965), Tarrant Maize: a new UK agricultural crop (1975) and Yapa The green revolution: a diffusion model (1977). Disease diffusion in agriculture has been an important area of study, particularly in Britain. Noted contributions came from Tinline Lee wave hypothesis for the initial pattern of spread during the

1967-68 foot-and-mouth epizootic (1972) and Gilg A study in agricultural disease diffusion: the case of the 1970-71 fowl-pest epidemic (1973). Both of these studies supported wave theory notions of spread rather than the development of multiple centres.

With the study of agricultural change came concern with means of change, particularly the development of extension services, the availability of information and problems of accessibility. Work in all these areas of study is widespread, particularly in the Third World. Three examples from India are Wanmali Rural service centres in India: present identification and the acceptance of extension (1975), Mayfield and Yapa Information fields in rural Mysore (1974) and Wilbanks Accessibility and technological change in Northern India (1972).

## Management and Decision Making

Concern with innovation diffusion and interest in decisions to adopt or not to adopt innovations has encouraged research into management aspects of agricultural geography at farm, corporation and government levels. Wolpert's (1964) seminal study has been cited already, as have studies of the influence of corporations. Continuing in the Wolpert vein is the work of Bowler (1975) cited above which identifies farms which deviate in enterprise specialization from the norm for their location. Farming is a high risk industry and risk and uncertainty are important factors in management decisions. Environmental risk was discussed by Saarinen Perception of the drought hazard on the Great Plains (1966), and game theoretic techniques for examining how farmers dealt with risk were applied by Gould Man against his environment: a game theoretic framework (1963) and Wheat on Kilimanjaro: the perception of choice within game and learning model frameworks (1965). Some critical comment on and discussion of these techniques may be found in Pred Behaviour and location vol. 1 (1967). Ilberry has tried to overcome some of the problems involved in his Point-score analysis: a methodological framework for analysing the decision-making process in agriculture (1977) whilst Chapman has examined the relationship of management to government in Perception and regulation: a case study of farmers in Bihar (1974). With management studies has come an appreciation of the importance of farmers' knowledge and perceptions and the importance of discovering and applying this knowledge to solve a variety of problems. Questionnaire interviews may often prove misleading. To avoid the dangers of the questionnaire in eliciting community environmental knowledge the repertory grid method has been applied notably by Townsend Farm "failures": the application of personal constructs in the tropical rainforest (1976) and Perceived worlds of the colonists of tropical rainforest, Colombia (1977) and by Richards Community environmental knowledge in African rural development (1979).

Government is a major factor in all agricultural activity and frequently a major determinant of spatial structures. Many of the papers cited above particularly the Third World studies of Chapman (1974), Townsend (1977) and Wanmali (1975), refer to the role of government. To these we may add Briggs *Farmers' responses to planned agricultural development in the Sudan* (1978). In the Western World extensive study has been made in North America, New Zealand and Europe. For example in the United States we have Fielding, *The Los Angeles milkshed: a study of the political factor in agriculture* (1964), Fisher *Federal crop allotment programme and responses by individual farmer operators* (1970) and Bailey, *Government policy and the location of cotton production* (1978). In New Zealand the very strong influence of government has been discussed by Fielding *The role of government in New Zealand wheat growing* (1965) and Stover *The government as farmer in New Zealand* (1969). In Britain Bowler has examined aid, subsidies and regional policy in *The adoption of grant aid in agriculture* (1976), *Spatial responses to agricultural subsidies in England and Wales* (1976) and *Regional agricultural policies: experience in the U.K.* (1976). In Europe Bowler has examined the effects of the Common Agricultural Policy on agriculture in *CAP and the space economy of agriculture in the EEC* (1976) and Tarrant has studied trade in *Agricultural trade within the European Community* (1980). International effects of rival producer policies and management of the international trade in food have been the subject of a major study by Tarrant *Food policies* (1980) which examines the gap in food production at international and regional levels, food crises, food aid and the scope of government intervention in agriculture. Tarrant examines food aid further in *The geography of food aid* (1980).

## Farming

Aspects of farming which have attracted the attention of geographers are structural characteristics including subdivision and fragmentation, farm size and farm organisation in relation to the other activities of farmers. The advantages and disadvantages of monoculture have been discussed by Brookfield *Problems of monoculture and diversification in a sugar island: Mauritius* (1959) and Igbozurike *Polyculture and monoculture: contrast and analysis* (1978). Subdivision and fragmentation have long been regarded as major problems of particular interest to geographers, especially in many countries where peasant farming is dominant and where holdings are very small. An important critique of the conventional view is Farmer *On not controlling subdivision in paddy-lands* (1960). In Europe and North America fragmentation is often caused by expansion in the size of farm holdings through purchase of properties which are not adjacent, but it can still create management problems. Useful discussion is in Smith *Fragmented farms in the United States* (1975) and Sublett *Farmers on the* road: interfarm migration and the farming of noncontinguous lands in three Midwestern townships, 1939-1969 (1975). Consolidation has also been the subject of geographical enquiry as in Lambert Farm consolidation in Western Europe (1963). Contributions to the technique of measuring fragmentation include Simmons An index of farm structure, with a Nottinghamshire example (1964) and Igbozurike Fragmentation in tropical agriculture: an overrated phenomenon (1970).

The use of machines, increasing efficiency in commercial operation and the interest in farming of workers and managers from industry have created other kinds of farming operation from the conventional idea of a whole-time unit with more than one enterprise which occupies the farmer and his labour force for the whole year. Kollmorgen and Jenks described variations from farming "norms" in Dakota and Montana Suitcase farming in Sully County, South Dakota (1958) and Sidewalk farming in Took County, Montana and Trail County, North Dakota (1958) Hobby farming has been described by Layton The operational structure of the hobby farm (1978) and part-time farming by Gasson Part-time farmers in south-east England (1966) and in Fuller and Mage (eds.) Part-time farming: problem or resource in rural development (1976).

Farm size has been of special interest and is often a potent factor in enterprise choice and distribution. Grigg The geography of farm size: a preliminary survey (1966) provides a general treatment and Gregor has examined the relationship of farm size to proximity to cities in The large farm as a stereotype: a look at the Pacific Southwest (1979). Of the various types of large farm which have attracted special interest the most widely treated are state and collective farms in the centrally planned economies and plantations. Symons has discussed state and collective farms in the USSR in Agricultural productivity and the changing roles of state and collective farms in the USSR (1966) and in Russian agriculture: a geographic approach (1972). On plantation agriculture the major work is Courtenay Plantation agriculture (1965) but valuable comment and discussion is in Gregor The changing plantation (1965) and Jackson Towards an understanding of plantation agriculture (1969).

The factors of agricultural production have received very varying treatment from geographers. Capital has been the subject of very little enquiry although important in the concept of intensity and evidently varying spatially. Labour has been of special interest partly because data have been fairly easy to obtain and partly because of the interest in standard man days as an input index noted earlier. Coppock examined the geographical distribution of requirements in Regional differences in labour requirements in England and Wales (1965) and Manley and Olmstead looked at labour inputs in Geographical patterns of labour input as related to output indexes of scale of operation in American agriculture (1965).

Gasson has examined Mobility of farm workers (1974) and produced a more general study Resources in agriculture: labour (1974). In the Third World labour studies have mostly been left to economists but the problem of intensification was studied by Eder Agricultural intensification and the returns to labour in the Philippine swidden system (1977). Land has attracted a more obvious interest and a variety of aspects has been treated. Land use has been the subject of numerous studies and considerable debate, particularly in Britain. Most of these studies have treated agriculture only in part. Nevertheless they do contain a great deal of information and ideas of relevance to agricultural geography. Here one may cite of special interest Stamp (1962) cited above, Wibberley and Edwards An agricultural land budget for Britain 1965-2000 (1971), Best The changing land use structure of Britain (1976) and Coleman Land use planning: success or failure? (1977). Work on land quality has already been cited. Land values and land ownership have been treated by Munton The state of the agricultural land market 1971-3: a survey of auctioneers' property transactions (1975) and Financial institutions: their ownership of agricultural land in Great Britain (1977). Land tenure and land reform aspects, mainly in Third World countries, have been summarised by King Land reform: a world survey (1977).

Physical constraints on farm production and a variety of ecological aspects are still abundantly treated despite the general emphasis by many geographers on economic and organisational studies. Much of the work on productivity has come chiefly from production ecologists and has been discussed amongst others by Simmons Ecology and land use (1966). Kovda The problem of biological and economic productivity of the earth's land areas (1971) has examined geographical aspects and Harris has reviewed The ecology of agricultural systems (1969). Taylor's The relation of crop distributions to the drift pattern in south-west Lancashire (1952) was an important study of the effect of soils on geographical distribution, whilst landform relationships have been studied by Hidore The relations between cash-grain farming and landforms (1963). Climatic factors have been the subject of widespread study as in Taylor Weather and agriculture (1967), Reid, Burge and Carr Climatic change and European agriculture (1978), Hewes Causes of wheat failure in the dry farming region, Central Great Plains, 1939-1957 (1965), Hoar Rainfall, rice yields and irrigation needs in West Bengal (1964) and Tinline Meteorological aspects of the spread of foot-and-mouth disease (1969). Problems of tropical agriculture have attracted a considerable number of research workers in geography. A useful general study is Grigg The harsh lands (1970) and amongst many papers one may cite Igbozurike Ecological balance in tropical agriculture (1971) and Eden Ecology and land development: the case of the Amazonian rainforest (1978).

Finally agricultural geographers are importantly concerned with the social aspects of farming, more especially with the ways in which social values affect decision making or new ways of living create new attitudes to farming. Williams The social study of family farming (1963) underlined the importance of sociological learning for agricultural geographers, emphasised by problems such as depopulation as in Szabo Depopulation of farms in relation to the economic conditions of agriculture on the Canadian Prairies (1965) or new developments as in Wilhelm The growth of the vacation farm as a rural settlement element in the Federal Republic of Germany (1977). Important contributions have come from Gasson Goals and values of farmers (1973) and Socioeconomic status and orientation to work: the case of farmers (1974). The immense cultural variety in the Third World and its evident relationship to crop choice and farming methods suggests an exploration of how this relationship affects geographical distribution, but such studies have mostly been left to anthropologists. A major study in geography has been Uhlig Hill tribes and rice farmers in the Himalayas and south-east Asia: problems of the social and ecological differentation of agricultural landscape types (1969).

REFERENCES

Adams, R.B. (1977) 'Changing crop combinations and agricultural exports: Upper Midwest response variability', Geogr. Perspectives 39, 8-16.

Allan, J.A. and Rosing, K.E. (1973) 'Disparities in the recognition of indicators of agricultural development - a Northwest Indian case.' Inst. Brit. Geogr. Dev. Areas Study Group symp.

Ashton, J. and Cracknell, B.E. (1960-61) 'Agricultural holdings and farm business structure in England and Wales'. J. of Agric. Econ. 14, 472-506.

Bailey, W.H. (1978) 'Government policy and the location of cotton production'. Southeastern Geogr. 18, 93-101.

Baker, O.E. (1926-32) 'Agricultural regions of North America'. Econ. Geogr. 2-8.

Barnes, F.A. (1958) 'The evolution of salient patterns of milk production and distribution in England and Wales'. Trans. Inst. Brit. Geogr. 25, 167-95.

Belshaw, D.G.R. and Jackson, B. (1966) 'Type-of-farm areas'. Trans. Inst. Brit. Geogr. 38, 89-93.

Bennett Jones, R. (1954) The pattern of farming in the East Midlands. Schl. of Agric., Univ. of Nottingham.

Best, R.H. (1976) 'The changing land use structure of Britain'. Town and Country Plan. 44, (3), 171-76.

Best, R.H. and Gasson, R.M. (1966) 'The changing location of intensive crops', Stud. in Rur. Ld. Use 6, Wye Coll.

Birch, J.W. (1954) 'Observations on the delimitation of farming-type regions, with special reference to the Isle of Man'. Trans. Inst. Brit. Geogr. 20, 141-58.

Birch, J.W. (1960) 'A note on the sample-farm survey and its use as a basis for generalized mapping'. Econ. Geogr. 36, 254-59.

Blaikie, P.M. (1971) 'Spatial organisation of agriculture in some north Indian villages'. Trans. Inst. Brit. Geogr.52, 1-40 (Pt. I) and 53, 15-30 (Pt. II).

Blaut, J.M. (1959) 'Microgeographic sampling: a quantitative approach to regional agricultural geography'. Econ. Geogr. 35, 79-88.

Boserup, E. (1965) The conditions of agricultural growth: the economics of agrarian change under population pressure. Allen and Unwin.

Bowden, L.W. (1965) The diffusion of the decision to irrigate. Dept. of Geogr. Univ. of Chicago, Res. Pap. 97.

Bowler, I.R. (1972) 'Co-operation: a note on governmental promotion of change in agriculture'. Area 4 (3),169-73.

Bowler, I.R. (1975) 'Factors affecting the trend to enterprise specialisation in agriculture: a case study in Wales'. Cambria 2 (2), 100-11.

Bowler, I.R. (1976) 'The adoption of grant aid in agriculture'. Trans. Inst. Brit. Geogr. N.S. 1 (2), 143-58.

Bowler, I.R. (1976) 'Spatial responses to agricultural subsidies in England and Wales'. Area 8 (3), 225-29.

Bowler, I.R. (1976) 'Regional agricultural policies: experience in the U.K.' Econ. Geogr. 52, 267-80.

Bowler, I.R. (1976) 'The C.A.P. and the space economy of agriculture in the E.E.C.' in R. Lee and P.E. Ogden, Economy and society in the E.E.C.:spatial perspectives. Saxon House.

Briggs, J.A. (1978) 'Farmers' responses to planned agricultural development in the Sudan'. Trans. Inst. Brit. Geogr. N.S. 3 (4), 464-75.

Bromley, R.J. (1971) 'Markets in the developing countries: a review'. Geogr. 56 (2), 124-32.

Brookfield, H.C. (1959) 'Problems of monoculture and diversification in a sugar island: Mauritius'. Econ. Geogr. 35, 25-40.

Buchanan, R.O. (1959) 'Some reflections on agricultural geography'. Geogr. 44 (1),1-13.

Carlyle, W.J. (1975) 'Livestock markets in Scotland'. Ann. Assoc. Amer. Geogr. 65, 449-60.

Carlyle, W.J. (1978) 'Store stock marketing by small firms in the crofting counties'. Scot. Geogr. Mag. 94, 113-23.

Chakravarti, A.K. (1973) 'Green revolution in India'. Ann. Assoc. Amer. Geogr. 63, 319-30.

Chapman, G.P. (1974) 'Perception and regulation: a case study of farmers in Bihar'. Trans. Inst. Brit. Geogr. 62, 71-93.

Chisholm, M. (1962, 3rd edn. 1979) Rural settlement and land use. Methuen.
Chisholm, M. (1964) 'Problems in the classification and use of farming-type regions'. Trans. Inst. Brit. Geogr. 35, 91-103.
Clark, G. and Gordon, D.S. (1980) 'Sampling for farm studies in geography'. Geogr. 65 (2),101-06.
Coleman, A. (1977) 'Land use planning: success or failure?' The Architects' Journal, 165 (3),94-134.
Coppock, J.T. (1960) 'The parish as a geographical-statistical unit'. Tijds. v. Econ. en Soc. Geogr. 51, 317-26.
Coppock, J.T. (1964) 'Crop-livestock and enterprise combinations in England and Wales'. Econ. Geogr. 40, 65-81.
Coppock, J.T. (1965) 'Regional differences in labour requirements in England and Wales'. Fm. Econ. 10, 386-90.
Coppock, J.T. (1965) 'The cartographic representation of British agricultural statistics'. Geogr. 50 (2),101-11.
Coppock, J.T. (1968) 'The geography of agriculture'. J. of Agric. Econ. 19, 153-75.
Coppock, J.T. (1964, rev. 1976) An agricultural atlas of England and Wales. Faber and Faber.
Coppock, J.T. (1971) An agricultural geography of Great Britain. Bell.
Coppock, J.T. (1976) An agricultural atlas of Scotland. John Donald.
Courtenay, P.P. (1965) Plantation agriculture. Bell.
Cruickshank, J.G. and Armstrong, W.J. (1971) 'Soil and agricultural land classification in County Londonderry'. Trans. Inst. Brit. Geogr. 53, 79-94.
Curry, L. (1963) 'Regional variation in the seasonal programming of livestock farms in New Zealand'. Econ. Geogr. 39, 95-118.
Curtis, L.F. (1978) 'Remote sensing systems for monitoring crops and vegetation'. Prog. in Phys. Geogr. 2 (1),55-79.
Dalton, R.T. (1971) 'Peas for freezing: a recent development in Lincolnshire agriculture. E. Midl. Geogr. 5, 133-41.
Datoo, B.A. (1973) Population density and agricultural systems in the Uluguru Mountains, Morogoro District. Bur. Resource Assess. and Land Use Plan., Univ. Dar-es-Salaam. Res. Pap. 26.
Datoo, B.A. (1978) 'Toward a reformulation of Boserup's theory of agricultural change. Econ. Geogr. 54 (2),135-44.
Dumont, R. (1957) Types of rural economy: studies in world agriculture. Methuen.
Eden, M.J. (1978) 'Ecology and land development: the case of Amazonian rain-forest. Trans. Inst. Brit. Geogr. N.S. 3 (4),444-63.
Eder, J.F. (1977) 'Agricultural intensification and the returns to labour in the Philippine swidden system'. Pac. Viewp. 18 (1),1-21.

Edwards, C.J.W. (1978) 'The effect of changing farm size upon levels of farm fragmentation: a Somerset case study'. J. Agric. Econ. 29, 143-53.
Edwards, C.J.W. (1980) 'Complexity and change in farm production systems: a Somerset case study'. Trans. Inst. Brit. Geogr. N.S. 5 (1) 45-52.
Erickson, F.C. (1948) 'The broken Cotton Belt'. Econ. Geogr. 24, 263-68.
Falkowski, J. (1977) An attempt at the typology and regionalization of agriculture in the Bydgoszcz-Torún agglomeration. Prz. Geogr. 49 (4) 713-30. (English summary).
Farmer, B.H. (1960) 'On not controlling subdivision in paddy lands'. Trans. Inst. Brit. Geogr. 28, 225-35.
Farmer, B.H. ed. (1977) Green Revolution? Technology and change in rice growing areas of Tamil Nadu and Sri Lanka. MacMillan.
Fielding, G.J. (1964) 'The Los Angeles milkshed: a study of the political factor in agriculture'. Geogr. Rev. 54, 1-12.
Fielding, G.J. (1965) 'The role of government in New Zealand wheat growing'. Ann. Assoc. Amer. Geogr. 55, 87-97.
Fisher, J.S. (1970) 'Federal crop allotment programme and responses by individual farmer operators'. Southeastern Geogr. 10 (2) 47-58.
Found, W.C. (1971) A theoretical approach to rural land-use patterns. Edw. Arnold.
Fuller, A.M. and Mage, J. eds.(1976) 'Part-time farming: problem or resource in rural development'. Geo Abstracts.
Garrison, W.L. and Marble, D.F. (1957) 'The spatial structure of agricultural activity'. Ann. Assoc. Amer. Geog. 47, 137-44.
Gasson, R.M. (1966) 'Part-time farmers in southeast England'. Fm. Econ. 11, 135-39.
Gasson, R.M. (1966) The influence of urbanisation in farm ownership and practice. Stud. in Rur. Ld. Use 7, Wye Coll.
Gasson, R.M. (1973) 'Goals and values of farmers'. J. Agric. Econ. 24 (3) 521-42.
Gasson, R.M. (1974) 'Socio-economic status and orientation to work: the case of farmers'. Soc. Rur. 14 (3) 127-41.
Gasson, R.M. (1974) Mobility of farm workers. Occ. Pap. 2. Dept. Ld. Econ. Univ. of Cambridge.
Gasson, R.M. (1974) 'Resources in agriculture: labour', in A. Edwards and A. Rogers, Agricultural Resources. Faber.
Geertz, C. (1963) Agricultural involution: the processes of ecological change in Indonesia. Univ. of Calif. P.
Gilg, A.W. (1973) 'A study in agricultural disease diffusion: the case of the 1970-71 fowl-pest epidemic'. Trans. Inst. Brit. Geog. 59, 77-97.
Gould, P.R. (1963) 'Man against his environment: a game theoretic framework'. Ann. Assoc. Amer. Geog. 53, 290-97.

Gould, P.R. (1965) 'Wheat on Kilimanjaro: the perception of choice within game and learning model frameworks'. Gen. Systems Yrbk 10, 157-66.

Gregor, H.F. (1965) 'The changing plantation'. Ann. Assoc. Amer. Geog. 55, 221-38.

Gregor, H.F. (1970) Geography of agriculture: themes in research. Prentice-Hall.

Gregor, H.F. (1979) 'The large farm as stereotype: a look at the Pacific Southwest'. Econ. Geogr. 55, 71-87.

Griffin, E. (1973) 'Testing the von Thünen theory in Uruguay', Geog. Rev. 63, 500-16.

Grigg, D.B. (1966) 'The geography of farm size: a preliminary survey'. Econ. Geog. 42, 204-35.

Grigg, D.B. (1969) 'The agricultural regions of the world: review and reflections', Econ. Geog. 45, 95-132.

Grigg, D.B. (1970) The harsh lands: a study in agricultural development. MacMillan.

Grigg, D.B. (1974) The agricultural systems of the world: an evolutionary approach. Cambridge Univ. P.

Griliches, Z. (1957) 'Hybrid corn: an exploration in the economics of technical change'. Econometrica 25, 501-22.

Griliches, Z. (1960) 'Hybrid corn and the economics of innovation.' Science 132, 275-80.

Hall, P. ed. (1966) Von Thünen's isolated state. Trans. C.M. Wartenberg. Pergamon.

Harris, D.R. (1969) 'The ecology of agricultural systems', in R.U. Cooke and J.H. Johnson, Trends in geography. Commonwealth and Internat. Library. Pergamon.

Hart, P.W. (1978) 'Geographical aspects of contract farming'. Tijds. v. Econ. en Soc. Geog. 69, 205-15.

Harvey, D.W. (1963) 'Locational change in the Kentish hops industry and the analysis of land use patterns'. Trans. Inst. Brit. Geog. 33, 123-44.

Harvey, D.W. (1966) 'Theoretical concepts and the analysis of land use patterns in geography'. Ann. Assoc. Amer. Geog. 56, 361-74.

Henshall, J.D. (1967) 'Models of agricultural activity', in R.J. Chorley and P. Haggett, Models in geography. Methuen.

Henshall, J.D. and King, L.J. (1966) 'Some structural characteristics of peasant agriculture in Barbados'. Econ. Geog. 42, 74-84.

Hervad-Jorgensen, K. (1977) 'Carrying capacity on Tikopia Island'. Geogr. Tids. 76, 88-95.

Hewes, L. (1965) 'Causes of wheat failure in the dry farming region, Central Great Plains, 1939-1957'. Econ. Geog. 41, 313-30.

Hidore, J.J. (1963) 'The relations between cash-grain farming and landforms'. Econ. Geog. 39, 84-89.

Hilton, N. (1962) 'A new approach to agricultural land classification for planning purposes', in Agric. Ld. Serv. Tech. Rep. 8, 91-102, Classification of agricultural land in

Britain.
Hilton, N. (1968) 'An approach to agricultural land classification in Great Britain'. Inst. Brit. Geog. Spec. Publ. 1. 127-44.
Hore, P.N. (1964) 'Rainfall, rice yields and irrigation needs in West Bengal'. Geogr. 49, 114-21.
Hunt, K.E. and Clark, K.R. (1966) The state of British agriculture 1965-1966. Agric. Econ. Res. Inst. Oxford.
Hunter, J.M. (1963) 'Cocoa migration and patterns of land ownership in the Densu Valley near Suhum, Ghana. Trans. Inst. Brit. Geogr. 33, 61-88.
Igbozurike, U.M. (1970) 'Fragmentation in tropical agriculture: an overrated phenomenon'. Prof. Geogr. 22, 321-25.
Igbozurike, U.M. (1971) 'Ecological balance in tropical agriculture'. Geogr. Rev. 61, 319-29.
Igbozurike, U.M. (1978) 'Polyculture and monoculture: contrast and analysis'. Geojournal 2, 443-49.
Ilberry, B.W. (1977) 'Point score analysis: a methodological framework for analysing the decision-making process in agriculture'. Tijds.v. Econ. en Soc. Geogr. 8, 66-71.
International Geographical Union (1965) The geographical typology of agriculture.Symposium in Agricultural Geography 1964. Liverpool Univ.
Jackson, J.C. (1969) 'Towards an understanding of plantation agriculture'. Area 4 (1) 36-41.
Jackson, R. (1972) 'A vicious circle? - consequences of von Thünen in tropical Africa'. Area 4 (4) 256-61.
Jensen, R.G. (1969) 'Regionalization and price zonation in Soviet agricultural planning'. Ann Assoc. Amer. Geogr. 59, 324-47.
Jonasson, O. (1925) 'Agricultural regions of Europe'. Econ. Geog. 1, 277-314.
Jones, G.E. (1963) 'The diffusion of agricultural innovations'. J. Agric. Econ. 15, 59-69.
Jones, G.E. (1967) 'The adoption and diffusion of agricultural practices'. Wld. Agric. Econ. Rur. Sec. Abs. Review Art. 6, 1-34.
King, R. (1977) Land reform. Bell.
Kollmorgen, W.M. and Jenks, G.F. (1958) 'Suitcase farming in Sully County, South Dakota'. Ann. Assoc. Amer. Geogr. 48, 27-40.
Kollmorgen, W.M. and Jenks, G.F. (1958) 'Sidewalk farming in Toole County, Montana and Trail County, North Dakota,' Ann. Assoc. Amer. Geogr. 48, 209-31
Kostrowicki, J. (1964) A geographical typology of agriculture: principles and methods. 20th Internat. Geogr. Cong. 220.
Kostrowicki, J. (1969) 'Agricultural typology'. Bull. I.G.U. 20, 36-40.
Kostrowicki, J. (1972) 'The typology of world agriculture: principles, methods and model types', in C. Vanzetti, Agricultural typology and land utilization. Proc. 4th Meet-

ing IGU Commission on Agric. Typol.
Kovda, V.A. (1971) 'The problem of biological and economic productivity of the Earth's land areas'. Sov. Geogr. 12, 6-23.
Krinks, P.A. (1978) 'Rural change in Java: an end to involution?' Geogr. 63 (1), 31-36.
Lambert, A. (1963) 'Farm consolidation in Western Europe'. Geogr. 48 (1) 31-48.
Layton, R.L. (1978) 'The operational structure of the hobby farm'. Area 10 (4) 242-46.
Leheron, R.B. and Warr, E.C.R. (1978) 'Corporate organisation, corporate strategy and agribusiness development in New Zealand'. N.Z. Geog. 32, 1-16.
Manley, V.P. and Olmstead, C.W. (1965) 'Geographic patterns of labour input as related to output indexes of scale of operation in American agriculture'. Ann. Assoc. Amer. Geog. 55 (4) 629-30.
Manshard, W. (1974) Tropical Agriculture. Longmans.
Mayfield, R.C. and Yapa, L.S. (1974) 'Information fields in rural Mysore'. Econ. Geog. 50, 313-23.
Morgan, W.B. (1978) Agriculture in the Third World: a spatial analysis. Bell and Hyman.
Morgan, W.B. and Munton, R.J.C. (1971) Agricultural geography. Methuen.
Munton, R.J.C. (1975) 'The state of the agricultural land market 1971-3: a survey of auctioneers' property transactions.' Oxf. Agrar. Stud. IV (N.S.) 2, 111-30.
Munton, R.J.C. (1977) 'Financial institutions: their ownership of agricultural land in Great Britain'. Area 9 (1) 29-37.
Munton, R.J.C. and Norris, J.M. (1969) 'The analysis of farm organisation: an approach to the classification of agricultural land in Britain'. Geogr. Annal. 52 B (2) 95-103.
Napolitan, L. and Brown, C.J. (1963) 'A type of farming classification of agricultural holdings in England and Wales according to enterprise patterns'. J. Agric. Econ. 15, 595-616.
O'Keefe, P. and Wisner, B. (1977) Landuse and development. IAI assoc. with Environment Training Programme UNEP - IDEP - SIDA. Special Report 5.
Parsons, J.W. (1977) 'Corporate farming in California'. Geogr. Rev. 67, 354-57.
Pred, A. (1967) 'Behaviour and location'. Vol.I. Lund. Stud. in Geogr. Ser. B., 27.
Preston, D.A. (1978) Farmers and towns: rural-urban relations in highland Bolivia. Geo Abstracts.
Reid, I.G., Burge, S.W. and Carr, M.K.V. (1978) 'Climatic change and European agriculture'. Weather 33, 193-95.
Richards, P. (1979) 'Community environmental knowledge in African rural development'. Inst. Dev. Stud. Bull. 10 (2)

28-36.

Rozlucki, W. (1977) 'The green revolution and the development of traditional agriculture: a case study of India'. Geogr. Polon. 35, 111-26.

Saarinen, T.F. (1966) Perception of the drought hazard on the Great Plains. Dep. Geog. Univ. of Chicago. Res. Pap. 106.

Sauer, C.O. (1952) Agricultural origins and dispersals. Amer. Geog. Soc. Bowman Mem. Lect. 2.

Schneider, R., Brown, L.A., Harvey, M.E. and Riddell, J.B. (1978) Innovation diffusion and development in a Third World setting: the case of the cooperative movement in Sierra Leone. Dep. Geog. Ohio State Univ. Disc. Pap. 54.

Simmons, A.J. (1964) 'An index of farm structure, with a Nottinghamshire example'. E. Midl. Geog. 3 (5) 255-61.

Simmons, I.G. (1966) 'Ecology and land use.' Trans. Inst. Brit. Geog. 38, 59-72.

Simpson, E.S. (1959) 'Milk production in England and Wales: a study in collective marketing'. Geog. Rev. 49, 95-111.

Sinclair, R. (1957) 'Von Thünen and urban sprawl'. Ann. Assoc. Amer. Geog. 47, 72-87.

Smith, E.G. (1975) 'Fragmented farms in the United States'. Ann. Assoc. Amer. Geog. 65, 58-70.

Smith, R.H.T. ed. (1978) Market place trade - periodic markets, hawkers and traders in Africa, Asia and Latin America. Centre Transp. Stud. Univ. of Birmingham.

Smith, W. (1974) 'Marketing-farm linkages and land use change: a Quebec case study'. Cah de Géogr. de Québec 18, 297-315.

Smith, W. (1975) 'The farm-market linkage: a note'. Area 7 (3) 180-2.

Spencer, J.E. and Horvarth, R.J. (1963) 'How does an agricultural region originate? Ann. Assoc. Amer. Geog. 53, 74-92.

Stamp, L.D. (1958) 'The measurement of land resources'. Geog. Rev. 48, 1-15.

Stamp, L.D. (1948, 3rd ed. 1962) The land of Britain: its use and misuse. Longmans.

Stola, W. (1977) 'An attempt to apply typological methods in a comparative study of agriculture in Belgium and Poland.' Prz. Geogr. 49, 757-71 (with English summary).

Stover, L. (1969) 'The government as farmer in New Zealand'. Econ. Geog. 45, 324-38.

Sublett, M.D. (1975) Farmers on the road: interfarm migration and the farming of noncontiguous lands in three Mid-Western townships, 1939-1969. Dep. Geog. Univ. Chicago. Res. Pap. 168.

Symons, L. (1966) 'Agricultural productivity and the changing roles of state and collective farms in the USSR.' Pac. Viewp. 7 (1) 54-66.

Symons, L. (1967) Agricultural geography. Bell.

Symons, L. (1972) Russian agriculture: a geographic approach. Bell.

Szabo, M.L. (1965) 'Depopulation of farms in relation to the economic conditions of agriculture on the Canadian Prairies'. Geog. Bull. 7, 187-202.

Tarrant, J.R. (1969) 'Some spatial variations in Irish agriculture'. Tijds. v. Econ. en Soc. Geog. 60, 228-37.

Tarrant, J.R. (1974) Agricultural geography. David and Charles.

Tarrant, J.R. (1975) 'Maize: a new U.K. agricultural crop.' Area 7 (3) 175-79.

Tarrant, J.R. (1980) 'Agricultural trade within the European Community'. Area 12 (1) 37-42.

Tarrant, J.R. (1980) Food policies. Wiley.

Tarrant, J.R. (1980) 'The geography of food aid'. Trans. Inst. Brit. Geog. N.S. 5 (2) 125-40.

Taylor, J.A. (1952) 'The relation of crop distributions to the drift pattern in south-west Lancashire'. Trans. Inst. Brit. Geog. 18, 77-91.

Taylor, J.A. (1967) Weather and agriculture. Pergamon.

Thomas, D. (1963) Agriculture in Wales during the Napoleonic Wars. Univ. of Wales.

Thomas, M.F. and Whittington, G.W. eds. (1969) Environment and land use in Africa. Methuen.

Thünen, J.H. von. (1826) Der Isolierte Staat in Beziehung auf Landwirtschaft und Nationalsökonomie. Part I. Perthes.

Tinline, R.R. (1972) 'Lee wave hypothesis for the initial pattern of spread during the 1967-68 foot-and-mouth epizootic', in N.D. McGlashan, Medical Geography: techniques and field studies. Methuen.

Tinline, R.R. (1969) 'Meteorological aspects of the spread of foot-and-mouth disease'. Biometeorology 4 (2) 102.

Townsend, J.G. (1976) 'Farm "failures": the application of personal constructs in the tropical rainforest'. Area 8 (3) 219-22.

Townsend, J.G. (1977) 'Perceived worlds of the colonists of tropical rainforest, Colombia'. Trans. Inst. Brit. Geog. N.S. 2 (4) 430-58.

Uhlig, H. (1969) 'Hill tribes and rice farmers in the Himalayas and south-east Asia: problems of the social and ecological differentiation of agricultural landscape types'. Trans. Inst. Brit. Geog. 47, 1-23.

Wanmali, S. (1975) 'Rural service centres in India: present identification and the acceptance of extension'. Area 7 (3) 169.

Watters, R.M. (1960) 'The nature of shifting cultivation'. Pac. Viewp. 1 (1) 59-99.

Watts, H.D. (1971) 'The location of the sugar beet industry in England and Wales, 1912-36'. Trans. Inst. Brit. Geog. 53, 95-116.

Weaver, J.C. (1954) 'Crop-combination regions in the Middle West'. Geogr. Rev. 44, 175-200.

Weaver, J.C. et al. (1956) 'Livestock units and combination

regions in the Middle West'. Econ. Geogr. 32, 237-59.

White, R.L. and Watts, H.D. (1977) 'The spatial evolution of an industry: the example of broiler production'. Trans. Inst. Brit. Geog. N.S. 2, 175-91.

Wibberley, G.P. and Edwards, A.M. (1971) An agricultural land budget for Britain 1965-2000. Stud. in Rur. Ld. Use 10. Wye Coll.

Wilbanks, T.J. (1972) 'Accessibility and technological change in Northern India'. Ann. Assoc. Amer. Geog. 62, 427-36.

Wilhelm, H.G.H. (1977) 'The growth of the vacation farm as a rural settlement element in the Federal Republic of Germany'. E. Lakes Geogr. 12, 1-10.

Williams, W.M. (1963) 'The social study of family farming'. Geogr. Journ. 129, 63-75.

Wolpert, J. (1964) 'The decision process in a spatial context'. Ann. Assoc. Amer. Geog. 54, 537-58.

Wood, W.F. (1955) 'Use of stratified random samples in land use study'. Ann. Assoc. Amer. Geog. 45, 350-67.

Yapa, L.S. (1977) 'The green revolution: a diffusion model'. Ann. Assoc. Amer. Geog. 67, 350-59.

Zvorykin, K.V. (1963) 'Scientific principles for an agro-production classification of lands'. Sov. Geog. 4, 3-9.

Chapter Five

# INDUSTRIAL GEOGRAPHY

Peter A. Wood

Of all the systematic branches of the subject, it is arguable that industrial geography has been transformed most profoundly in its focus and emphasis during the past 15-20 years. In the early 1960s, theoretical discussion was still dominated by the 'classical' analysis of optimum location patterns, relying heavily on Alfred Weber's Theory of the Location of Industries (1909, Translated by C.J. Friedrich, University of Chicago Press, 1929) and August Lösch's The Economics of Location (1944, Translated by W.H. Woglom, Yale University Press, 1954). Weber's analysis explained where the minimum costs of production could be achieved by different industries, assuming profit maximising goals by perfectly competitive firms, operated by completely knowledgeable entrepreneurs. Lösch, seeking more general principles of rational location, analysed the market areas that would be required to support various industries in a system of cities. In the mid 1950s, Walter Isard added his development of these and other earlier contributions in Location and Space Economy (M.I.T. Press, 1956).

Empirical work, on the other hand, remained primarily descriptive, orientated towards historical accounts of particular industries. Some studies also examined why industrialists chose to locate in certain areas (for example see W.F. Luttrell, Factory Location and Industrial Movement N.I.E.S.R., 1962). The lack of correspondence between this evidence and the processes postulated by classical theory was striking and also damaging to the credibility of the subject. 'Psychic income' and 'satisfying behaviour' were introduced to explain the deviations of practice from theory (see M.L. Greenhut, Plant Location in Theory and Practice, University of North Carolina Press, 1956) and the term 'behavioural' entered the literature from economics (R.M. Cyert and J.G. March, A Behavioural Theory of the Firm, Prentice Hall, 1963). D.M. Smith developed the notion of the 'spatial margins of profitability', to define an area around a theoretically optimum location within which firms can receive satisfactory profits and thus survive. The 'behavioural matrix' was devised by A.R. Pred to explain deviations from optimum locational behaviour in terms of industrialists' varying access to information and ability to use it effectively. (D.M. Smith, Industrial Location, Wiley, 1971, Second Edition, 1981; A.R. Pred, Behaviour and Location, Lund Studies in Geography, Series B, No.27, 1967; No.28, 1969).

Although Pred was severely critical of the simplified 'economic man' assumptions of classical location theory and attempted to incorporate the complexity of location decision-making into his matrix, the industrial location problem has generally come to be reviewed since in a somewhat different perspective. In essence, modern industrial geography is less preoccupied with the spatial pattern of manufacturing output in itself and more with the broader spatial ramifications of the organisation of production. Which developments in the structure of manufacturing at large are likely to result in differential changes in activity from place to place? Why do communities in different locations tend to specialise both in the goods they produce, and increasingly in the specific manufacturing functions they perform? The answers to these questions are sought through studies of the organisation of production by different types of firm responding to various external changes; large and small firms, publicly or privately owned, locally, nationally or internationally controlled, involved in making single products or a wide range of output, in traditional or innovatory technologies. It is assumed that the effects of wider national and international changes upon local economies can only be understood by placing individual plants in their corporate context.

This perspective, of course, arises from a somewhat belated recognition by industrial geographers of the dominant importance in modern capitalism of large multi-plant firms, whose internal processes of resource allocation may be as significant for location change as the conventionally recognised processes of technological and market change. The tradition of local industrial ownership and control which evolved during the eighteenth and nineteenth centuries has been in decline for at least the past fifty years, along with the specialised industrial regions which it supported. In both the surviving elements of the old industries and the new industries of the mid-late twentieth century, regional economies have increasingly become dominated by organisations whose centres of decision-making are located elsewhere, whether in publicly or privately-owned corporations. Small-medium sized firms retain a significant role within most industrial regions, of course, although their relative decline has become a cause for alarm, but the processes and products in which they can compete are increasingly determined by the policies of large organisations.

A further reason for interest in the corporate context of industrial location change has been the expansion of government efforts to influence location decisions. In Britain, in particular, something like £700 million per year was being spent in the mid 1970s on regional policy, to attract investment to various types of Assisted Area. Was this investment justified? Could it be more closely tailored to the operational needs of different types of company? How could the attractions of different areas be presented to industrialists? What was the relationship of this area-specific aid to the support given to particular industries? Were company strategies being significantly affected by the incentives, or were they merely exploiting government regional funds to subsidize investments that would have taken place in any case? In an age of intensive planning activity, an understanding of the locational priorities and motivations of corporate decision-makers was obviously important.

The rest of this chapter will emphasize the more recent literature on the spatial impacts of industrial corporate structure and change. It will also concentrate on British experience, even though important conceptual and empirical contributions have been made in North America, for example on the regional impacts of technological change, and in Scandinavia, where the relationship between corporate organisaton and regional development has been most thoroughly explored. In the space available this bias should allow the contribution of industrial geography to be examined in relation to the distinctive problems of a single industrial nation. It will also enable a cumulative picture of late twentieth century conditions to emerge from the various threads of enquiry. In referring to particular contributions, especially in the form of periodical articles, the most recent papers will be cited, on the assumption that the earlier literature can be traced through their reference lists.

## Types of Literature on Industrial Geography

In view of the rapidity and novelty of developments in the subject during the past decade most of the up-to-date literature on industrial geography is to be found in periodical articles. Many journals are involved, of course, and a good many ideas from non-geographical sources have been incorporated into recent writings. Some journals, however, can be regarded as providing the specialist base of the subject. Regional Studies, for example, has almost become the 'house journal' of British industrial geography, and in fact published a special issue containing a valuable set of representative papers on 'Organisation and Industrial Location in the United Kingdom' in 1978 (Vol.12 (2)). Another useful journal which recently devoted an issue to 'Multinational Companies' (Vol.71 (1), 1980), is the Dutch based, although predominantly English language, Tijdschrift voor Economische en Sociale Geografie. Progress in Human Geography includes an annual progress review on industrial geography, written by David Keeble between 1977 and 1979 (Vol. 1, 304-12; Vol. 2, 318-23; Vol. 3, 425-33) and by the present author in 1980 (Vol. 4, 406-16). Urban Studies has regularly published papers on urban industrial change, especially reflecting recent concern about the decline of inner city manufacturing, and Environment and Planning (A) also contains some important contributions on industrial location changes. Finally, the Institute of British Geographers' Area has included some lively exchanges over the years on such topics as the impact of regional policy in Britain and the scope and relevance of industrial linkage studies.

There is at present no publication that could be regarded as a representative or comprehensive textbook of modern industrial geography, although the relevant chapters in P. Lloyd and P.E. Dickens's, Location in Space (2nd edn. Harper and Row, 1977, Chapters 4-10) are as good a general introduction as any. Edited compilations of papers have formed the basis of several useful volumes in recent years, although the subject matter and quality of individual contributions vary and the editors seldom seem able to bring them together very convincingly around a coherent theme. Two important such books were published in 1974/5, giving the sub-discipline a considerable fillip as a result; F.E.I. Hamilton's Spatial Perspectives on Industrial Organisation and Decision Making (Wiley, 1974) and L. Collins and D.F. Walker's Locational Dynamics of Manufacturing Activity (Wiley, 1975). Hamilton also edited two further volumes which were published in 1978, based on papers presented to a meeting of the International Geographical Union Commission on Industrial Systems; Contemporary Industrialisation and Industrial Change (both Longmans,1978). In 1979 Hamilton, with G. Linge edited another volume, Spatial Analysis, Industry and the Industrial Environment, Volume 1; Industrial Systems, intended to be the first of a series of collections reflecting international developments in industrial location studies. Most recently, in 1980, D.F. Walker has edited a set of essays, incorporating international comparison of development experience, Planning Industrial Development (Wiley, 1980). All of these books contain useful chapters on aspects of industrial location change in the United Kingdom, but their main value, judged in the present context, is to place such developments against the background of international enquiry.

Finally in this section, reference should be made to the published literature on geographical developments in particular industries in Britain. Of course, industrial economists and economic historians have published more regularly on such themes as the problems of the vehicle industry eg. P.J.S. Dunnett, The Decline of the British Motor Industry (Croom Helm, 1980) and K. Bhaskar, The Future of the UK Motor Industry

Table 1
Guide to Research Studies on the Industrial Geography of Britain

| Scale: | Aggregate Studies (Secondary Data) | Firm ('micro-') level Studies (Primary Data) |
|---|---|---|
| <u>THEME</u><br>Causes for location change | 1<br>(a) National trends in industrial location change<br>(b) Regional and sub-regional patterns of: i) manufacturing employment and structure ii) Components of change (Establishment-level data) | 2<br>Decision processes leading to location change within firms<br>Plant movement, and the influence of policy<br>Corporate rationalisation, including merger and takeover activity. |
| Patterns of Organisation: materials linkages and information contacts | 3<br>(a) Materials exchange: Input-output analysis and patterns of industrial localization.<br>(b) Evidence for extended patterns of information exchange: Regional patterns of<br>i) Ownership<br>ii) Headquarterlocations<br>iii) R & D facilities<br>iv) Foreign-owned firms 'External control and regional development' | 4<br>Patterns of inter-plant organisation: Theoretical discussions.<br>Empirical studies, including multiplier effects of material and information exchange patterns. |

(Kogan Page, 1979). Three particular contributions with a specifically geographical bias are worth note, however; K. Warren's study of The British Iron and Steel Industry since 1840 (Bell, 1970), D.W. Heal's examination of more recent events in The Steel Industry in Post-War Britain (David and Charles, 1974) and K. Chapman's North Sea Oil and Gas (David and Charles, 1976).

## Key Themes in Industrial Geography

Two themes form the basis of modern industrial geography and are reflected in its burgeoning literature on the United Kingdom. The first is the medium-long term pattern of change in the location of industrial investment and employment, and its causes. Such change, of course, is the result of many influences upon manufacturing and commercial companies, acting as intermediaries between general macro-economic trends and their geographical outcomes. The second theme is the current spatial pattern of organisation between establishments, within and between firms, exchanging raw, semi-finished or finished materials and routine and more strategically significant information. A corollory of such 'linkage' patterns is the impact or 'multiplier effect' of operational changes at one location upon others, whether locally or farther afield. These two themes encompass many detailed facets of industrial activity at different locations which have been given attention by industrial geographers. They have generally been examined at two scales; the aggregate level, concerned with broad patterns of national or regional activity, involving many companies, and the individual or 'micro-' level, at which the focus is upon the behaviour of particular firms or groups of firms. These two levels of generalisation reflect to a considerable degree complementary approaches to the collection and analysis of data. Aggregate studies rely predominantly upon secondary sources, mainly from government departments, and for this reason usually measure activity in terms of employment. Insights into firm behaviour, on the other hand, are usually based upon the primary collection of various information about their operations from firms themselves, through questionnaire surveys. Table 1 indicates a fourfold classification of work in industrial geography, upon which this account will be based, and summarizes the detailed questions which arise under each category.

### 1 Aggregate Studies of Patterns of Location Change

#### (a) National Trends

Analysis of aggregate trends in the location of British manufacturing employment has been undertaken both on the national pattern and for individual regions. In both cases the nature of the secondary data employed has largely determined the form of analysis and the types of conclusions achieved. For example, the study by M. Chisholm and J. Oeppen, The Changing Pattern of Employment (Croom Helm, 1973) was based on 1959 and 1968 information on 152 industrial categories for 61 planning sub-regions, supplied by the Department of Employment. As the authors admit, many intriguing questions were left unanswered as a result of dependence upon a single type of data. They were, however, able to identify significant trends towards the diversification of formerly specialised areas and the wider dispersion of employment in many manufacturing industries. More generally, they noted a convergence of sub-regional employment structures towards the national average. One of the few studies which did not depend upon employment information for its analysis of aggregate patterns was Freight Flows and Spatial Aspects of the British Economy (Cambridge, 1973) by M. Chisholm and P. O'Sullivan. This employed a Ministry of Transport data set for 1964 which traced the volumes of freight flowing between 78 zones in Great Britain for 13 commodities. Although such an exercise has not been repeated since, this study provided tantalizing insights into the pattern of sub-regional exchange which forms the basis of the British manufacturing space economy. A shorter account of this work is to be found in Spatial Policy Problems of the British Economy (Cambridge, 1971), edited by M. Chisholm and G. Manners, which also contains valuable chapters on employment mobility by D. Keeble, the British energy market by G. Manners and the chemical industry by K. Warren.

The most accessible and up-to-date introduction to the general pattern of change in British employment, manufacturing, energy supply, transport and planning is to be found in J.W. House (Ed.), The U.K. Space Wiedenfeld and Nicholson, 2nd. edn. 1977). Probably the most comprehensive single review of trends in post-war British industrial geography, however, is Industrial Location and Planning in the United Kingdom by David Keeble (Methuen, 1976). The key chapters (2-5) are based upon mapping and multivariate regression analysis of sub-regional manufacturing employment and floorspace change between 1959, 1966 and 1971. This demonstrates an important change that appears to have taken place during the 1960s. The division accepted by regional policy, between the 'core' of growth and prosperity in and around the London and West Midland conurbations and the 'periphery' of the northern and western regions, seemed to become less dominant. After the mid 1960s, the trend of manufacturing employment change in certain respects reversed, and is summarized by Keeble in terms of a 'periphery-centre' model in which industry had begun to decline rapidly in all the large conurbations, including London and the West Midlands, and grow only in rural areas and small-medium sized towns. Keeble extended his analysis up to 1976 in 'Industrial decline, regional policy and the urban-rural manufacturing shift in the United Kingdom'

(Environment and Planning (A), 12 (8), 1980, 945-62), and his conclusions have been confirmed by the work of S. Fothergill and G. Gudgin, in Regional Employment Change: A subregional explanation (Progress in Planning, 12, 155-219, Pergamon, 1979). This monograph examines the longest available time series of industrial employment data, from 1952 to 1975, and traces the increasing tendency since 1966 for contrasts in employment change within regions to become more significant than those between regions, especially at the expense of the conurbations and large cities.

These new spatial patterns have emerged against a background of decline in total manufacturing employment, in contrast to the earlier growth, and the statistical approach to spatial change has been criticized for its detachment from consideration of trends in the economy at large. An important element in recent change, for example, but neglected in most interpretations of spatial employment patterns, has been the tendency for many firms under competitive pressure, to increase productivity, shedding labour while maintaining output. D. Massey and R. Meegan have identified the variable effects on employment of productivity, as distinct from output changes in different regions between 1968 and 1973 ('Labour productivity and regional employment change' Area 11 (2), 1979, 137-45).

The search for explanations of patterns of location change in terms of the attributes of places, rather than as by-products of 'structural' changes in the organisation of manufacturing at large, has to some extent been influenced by the importance of regional policy in Britain, attempting to manipulate the attractiveness of different places and regions for industrial investment. As well as in Keeble (1976, Chapters 8 and 9), the history and purpose of regional industrial planning in Britain are fully explored in G. McCrone's Regional Policy in Britain (Allen and Unwin, 1969) and in Regional Development in Britain, edited by G. Manners (Wiley, 2nd edn., 1980). B. Moore and J. Rhodes published a series of papers in the mid-1970s, based on their attempts to quantify the employment created by regional policy in development areas, especially during its period of greatest activity in the late 1960s and early 1970s (See chapters in A. Whiting (Ed.) The Economics of Industrial Subsidies, HMSO 1976; M. Sant (Ed.), Regional Policy and Planning for Europe, (Saxon House), 1974; and J. Vaizey (Ed.) Economic Sovereignty and Regional Policy (Gill and Macmillan, 1975). Their estimates have been subject to criticism and revision by R.R. Mackay (also represented in Whiting's volume) and B. Ashcroft and J. Taylor ('The movement of manufacturing industry and the effects of regional policy', Oxford Economic Papers, 29 (1), 1977, 84-101). J.A. Schofield has written a useful review of some of the technicalities of these and other exercises in, 'Macro-evaluations of the impact of regional policy in Britain: a review of recent research'

(Urban Studies, 16, 1979, 251-71).

Empirical examination of the patterns of industrial movement encouraged by regional policy has been greatly assisted through the publication by the Board of Trade of a study by R.S. Howard, The Movement of Manufacturing Industry in the United Kingdom, 1945-65 (HMSO, 1968). This and later data from the same source have been further analysed both by Keeble (1976, Chapter 6) and by M. Sant, in Industrial Movement and Regional Development: the British Case (Pergamon, 1975).

(b) Regional Trends

Aggregate studies of industrial trends within particular regions have taken two forms, one predominating in the 1960s, the other gaining significance during the 1970s. In the earlier period local employment data from Censuses (1951, 1961, 1966 (Sample) and 1971) and from the Ministry of Labour (Later the Department of Employment) usually provided the basis for the analysis of comparative industrial trends at the local level. Augmented by detailed local knowledge on particular industries, transportation developments, energy supply changes and the effects of planning policies, this approach was typified by the valuable 'Industrial Britain' series published by David and Charles; The North West, by D.M. Smith (1969), The North East, by J.W. House (1969), The Humberside Region, by P. Lewis and P. Jones (1970), South Wales, by G. Humphrys (1972) and The West Midlands by P.A. Wood (1976). Other regional industrial studies included J. Britton's book on the Bristol region, Regional Analysis and Economic Geography (Bell, 1967), and K. Warren's and J.M. Hall's contributions to the 'Problem Regions of Europe' series on North East England (Oxford, 1973) and London (Oxford, 1976) respectively.

During the early 1970s, many regional studies were undertaken by planners and consultants as a result of the enthusiasm for regional planning at that time, although generally these were weakest in the treatment of the economic base of regions. Data sources also changed, with the introduction of the Annual Census of Employment in 1972 to replace the often inaccurate tabulations derived from the annual sample count by the Department of Employment. The ACE survey, although probably more reliable, was slow to be disseminated because of technical difficulties and was also curtailed as part of government economies in the late 1970s. These data changes made the established form of local economic study very difficult to continue during the 1970s.

At the same time new ideas about local economic monitoring were evolving under the influence of the micro-level studies of manufacturing change then being widely undertaken. Even in the 1960s, some studies had used data sources which allowed employment changes in individual manufacturing plants, to be monitored, including openings and closures. In 1968,

P.E. Lloyd and P. Dicken pointed out the deficiencies of Census and Ministry of Labour information. They also drew attention to the possibility of employing computer-based data banks to trace the changing fortunes of local economies, using establishment-level employment data from the records of H.M. Factory Inspectorate and the Ministry of Labour ('The data bank in regional studies of industry'. Town Planning Review, 38, 1968, 304-16).

In the 1970s, a few such data banks for some urban regions came into being through the sterling efforts of individual researchers. The best known, in terms of published research results, was created by Lloyd and Dicken themselves at Manchester University, for the North West region. This provided location-specific employment information for all plants in Merseyside for 1959, 1966, 1972 and 1975, in Greater Manchester for all but the first of these dates, and in the rest of the North West for 1974/75. The basic source, the records of the Factory Inspectorate, was augmented to give a record which identified plant births, deaths, and transfers within the region, in situ employment change, and ownership links. Another region well served by detailed monitoring of manufacturing change has been the Clydeside conurbation. The Glasgow University Register of Industrial Establishments (GURIE) is based upon Department of Employment records, again augmented by local research, especially on plant ownership. In its original form the Register contained information for 1958 and 1968, but it has since been updated to 1973. Other plant level registers of local employment change have been accumulated, usually from augmented Factory Inspectorate data, for the East Midlands by G. Gudgin and J.J. Fagg, and for Stoke on Trent by J. Whitelegg and A. Moyes (J. Whitelegg,'Births and deaths of firms in the inner city' Urban Studies, 13 (3), 333-8). In Greater London, the Department of Industry has maintained a register of openings and closures of factories employing 20 or more people since 1966.

These attempts to set up establishment-based data banks culminated in the mid 1970s in a series of 'components of change' studies. In these, manufacturing employment change for various areas was categorized according to whether it had resulted from the closure or opening of plants, plant movement, or from in situ change of employment in surviving plants. In some cases, changes in the organisational status of local plants have also been systematically examined. Valuable insights were thus gained into the nature of urban manufacturing change in Manchester and Merseyside, Clydeside, the West Midlands, London and the East Midlands (P. Dicken and P.E. Lloyd, 'Inner metropolitan industrial change, enterprise structures and policy issues; case studies of Manchester and Merseyside' Regional Studies, 12 (2), 1978, 181-98; P.E. Lloyd, 'The components of industrial change for the Merseyside inner area: 1966-75' Urban Studies, 16, 1979, 45-60; C. Mason, 'Industrial decline in Greater Manchester, 1966-75; a components of change approach' Urban Studies, 17 (2), 1980, 173-84; J.R. Firn and J.K. Swales, 'The formation of new manufacturing establishments in the central Clydeside and West Midlands conurbations, 1963-1972: a comparative analysis' Regional Studies, 12 (2), 199-214; R. Dennis, 'The decline of manufacturing employment in Greater London, 1969-1974', Urban Studies, 15 (1), 63-74; G.H. Gudgin, Industrial Location Processes and Regional Employment Growth, (Saxon House, 1978).

These studies offer detailed exemplification of the diversity of processes underlying the aggregate trends identified by Keeble and by Fothergill and Gudgin, and especially how inner city areas have become unattractive to manufacturing enterprise. Plant closures had provided by far the major contribution to employment decline there in all cases. Most obviously, in Merseyside and in Newcastle, Birmingham, West Yorkshire and London's East End (Community Development Project, The Costs of Industrial Change, CDP Inter-Project Editorial Team, 1977), multi-plant firms were closing their inner city and old dockland plants and focusing investment elsewhere. As in Clydeside, new investment was being directed to the outer metropolitan areas and beyond, often in new towns. In Manchester and the West Midlands, the survival problems of locally based small firms were particularly significant, while the East Midlands, at least, with its constellation of medium sized cities and towns, still offered a suitable environment for seedbed growth (J.J. Fagg, 'A re-examination of the incubator hypothesis; a case study of Greater Leicester' Urban Studies, 17 (1), 1980, 35-44). As Fothergill and Gudgin suggested as a result of their study of national trends, local circumstances in various areas seemed to be determining the fortunes of the urban economies of Britain, and regional policy in the mid 1970s was much less in tune with this trend of events than in the 1960s.

## 2. Studies of the Processes of Location Change in Firms

The stimulus to behavioural studies of industrial location change in the 1970s was derived not only from a recognition that local change has various components, but also that within individual firms, plant openings, closures and *in situ* changes result from an extended chain of decision, usually arising from business considerations which have little to do with location *per se*. Individual firms usually set general goals concerning the products they sell, their levels of output, sales strategies, and policies towards technological innovation or employment, which result in particular investment decisions. In some cases such investment may be possible within existing plants, thus changing their function. On the other hand, plant expansion, contraction or closure may be necessary at various sites, or a new factory may even be required. The takeover of other firms might also be an appropriate response, so that their production facilities become integrated into the operational pattern of the acquiring firm. Perhaps the most comprehensive survey of the various types of location change carried out by firms has been provided by F.E.I. Hamilton, in 'Aspects of industrial mobility in the British economy' (Regional Studies, 12 (2), 1978, 153-66). In his sample of 1486 firms operating between 1960 and 1972, no fewer than 944 had set up at least one new plant, 400 had established more than one, and 354 firms had relocated plants. On the other hand, 617 firms had closed 724 plants. In many cases, of course, the same firm had undertaken both openings and closures, as well as other types of locational adjustment. For example, 43 per cent of the sample firms had experienced spatial changes in their operations as a result of merger or takeover activity, while all the firms working at more than one location had adjusted the roles and relative importance of established units, including in situ expansion in the cases of 1000 firms.

The processes behind this level of locational adjustment by manufacturing firms were explored in two important studies published in the early 1970s. The first, by P.M. Townroe, Industrial Location Decision; A Study in Management Behaviour (University of Birmingham, Centre for Urban and Regional Studies, Occasional Paper No. 15, 1971) examined the sequence of decision, which leads to plant movement. The second, by D.J. North and based on a study of the plastics industry, examined all types of location change. ('The process of locational change in different manufacturing organisations' in F.E.I. Hamilton, Spatial Perspectives on Industrial Organisation and Decision-Making, Wiley, 1974). As with aggregate studies of employment change, the focus of enquiry into firm behaviour in the early 1970s was influenced by the need to monitor the effects of regional policy incentives. This explains the emphasis of a number of studies on the reasons for industrial movement and its regional impacts. Townroe's later work investigating firms moving to East Anglia and Northern England ('Branch plants and regional development', Town Planning Review, 46 (1), 1975, 47-62 and a number of earlier papers), D.J. Spooner's enquiry into firms moving to the South West ('Some qualitative aspects of industrial movement in a problem area of the United Kingdom' Town Planning Review, 45 (1), 1974, 63-83) and the Department of Trade and Industry's own comprehensive survey of the effects of regional incentives on industrial movement, provide the main examples (Trade and Industry sub-committee of the House of Commons Expenditure Committee, Minutes of Evidence, Session 1972-3, (1973), 525-668). The effects of regional policy on firms in the West Midlands and North West are examined in M.J.M. Cooper's 'Government influence on industrial location: case studies in the North West and West Midlands of England' (Town Planning

Review, 47 (4), 1976, 384-97), and D.H. Green has published an interesting study of the imperfect knowledge of regional policy incentives held by industrial decision-makers ('Industrialists' information levels of regional incentives' Regional Studies, 11 (1), 1977, 7-18). Also emphasizing the point of view of the industrialist and the practical problems of establishing new factories, Townroe has written Planning Industrial Location (Leonard Hill, 1976).

Broader issues concerning patterns of locational adjustment by firms have been taken up by H.D. Watts in The Large Industrial Enterprise (Croom Helm, 1980), and in a series of earlier papers which consider the locational results of the rationalisation of production and distribution by large firms. The spatial impacts of takeover and merger activity have also been recently investigated by R. Leigh and D.J. North ('Regional aspects of acquisition activity in British manufacturing industry' Regional Studies, 12 (2), 1978, 227-46, and a chapter in F.E.I. Hamilton's Contemporary Industrialisation, Wiley, 1978). An important study by D. Massey and R. Meegan, The Geography of Industrial Reorganisation (Progress in Planning, 10, 155-237, Pergamon, 1979) attempts to establish a direct connection between the local employment impacts of mergers and the macro-economic conditions which caused them, based upon the example of the government-aided rationalisation of the electrical engineering industry between 1968 and 1972. A wider debate is thus now taking place about how aggregate location trends in British manufacturing industry can be explained through clearer theorisation of the ways in which individual firms respond to macro-economic changes, particularly in circumstances of recession. The preoccupation, understandably in the 1980s, is thus more with retrenchment and rationalisation, than with investment and growth. The prospect of accelerated future change as a result of technical innovations in electronics has added greater urgency to these issues.

## 3. Aggregate Studies of Industrial Linkage and Contact Patterns

### (a) Materials linkage and industrial location

One of the most persistent features of the geography of manufacturing is the tendency for different activities to gather together in particular regions. This agglomeration, or 'swarming' has for long been explained by the exchange of materials, or 'linkage' between industries which is a necessity of manufacturing; what Alfred Weber himself termed, 'the social nature of production'. Costs can be reduced in such agglomerations through various external economies of scale which result from the bulk supplying of materials, equipment and finished goods by specialist merchants, and the mass provision of transportation facilities and public services. The benefits of operating in large industrial regions also extend to the exchange of non-material, information resources. Skilled managerial expertise, and specialist knowledge about technical and market changes are supposedly more easily come by in established industrial areas compared with elsewhere.

The main difficulty with this 'linkage' explanation of industrial concentration is that little direct measurement of the aggregate importance of linkage has ever been made. In addition, as we have seen already, the attractiveness of the large established industrial agglomerations seems to be on the wane, perhaps as a result of the declining influence of linkage and external economies of scale on location patterns. Cheaper transportation and better communications would certainly explain the release of manufacturing from its ties to the nineteenth century conurbations. On the other hand, although traditional industrial regions are becoming more diversified and dispersed in their manufacturing activities, new forms of regional specialisation appear to be emerging, most startlingly through the concentration of 'information-processing' functions into the South East. Even if the significance of industrial linkage for location can be established therefore, its character and consequences appear to have been changing in recent decades. Consequently, in examining general trends, material exchanges between <u>industries</u> must be distinguished from information exchanges between different <u>functions</u> within large firms, such as production, administration or research and development. The latter may, of course, differentiate between sections of the same industry.

The best aggregate evidence for the remaining influence of material linkages on industrial distributions is to be found in W.F. Lever's demonstration that industries which transacted a high proportion of their business with each other at the <u>national</u> scale (as evidenced by the value of exchange between them in the 1963 Input-Output tables), tended to have similar location patterns. ('Industrial movement, spatial association and functional linkage' <u>Regional Studies</u>, 6, 1972, 371-84). Later, he related patterns of industrial movement between 1945 and 1965 to sub-regional industrial structures and demonstrated that, with the exception of the West Midlands, regions with a predominance of industries which were interconnected nationally tended to contribute least to inter-regional patterns of industrial mobility ('Mobile industry and levels of integration in sub-regional economic structures' <u>Regional Studies</u>, 9, 1975, 265-78). In the 1960s at least, therefore, the material inter-dependence of different industries in Britain seemed still to be significant in explaining their aggregate location patterns and behaviour.

(b) Information exchange: the evidence of ownership patterns

Of course, Lever's studies did not measure inter-industry materials linkage within regions directly. Evidence for the aggregate influence of information links on location change has been similarly indirect, inferred largely from tracing the changing geographical patterns of manufacturing ownership. As large multi-plant firms have grown in national importance, local and regional economies have come to depend more upon factories owned and, thus it is assumed, mainly controlled from outside. Complex patterns of control and information exchange spreading far beyond the local environment are regarded as increasingly crucial to local economic fortunes. Studies of the locations of large company headquarters (see J.B. Goddard and I.J. Smith, 'Changes in corporate control in the British urban system, 1972-77' Environment and Planning (A) 9, 1978, 1073-84), of research and development activities (R.J. Buswell and E.W. Lewis, 'The geographical distribution of industrial research activity in the U.K.' Regional Studies, 4, 1970, 297-306) and of other high order functions (G.F. Parsons, 'The giant manufacturing corporations and regional development in Britain' Area, 4, 1972, 99-103) all reveal a relatively high and growing concentration of control activities into the 'information rich' South East (J. Westaway, 'The spatial hierarchy of business organisation and its implications for the British urban system' Regional Studies, 8, 1974, 145-55). Particular alarm has been expressed at these developments from the point of view of individual regions, by geographers and economic planners. J. Firn's influential paper, with obvious political connotations, 'External control and regional development: the case of Scotland' (Environment and Planning (A), 7, 1975, 393-414), traced the declining importance of locally owned manufacturing in Scotland during the post war period, and the growth of external control, especially from South East England. He also examined the consequences of these trends for the Scottish economy and the reasons for the apparently low levels of indigenous enterprise there. Similar patterns of change have almost certainly affected all other regions. I.J. Smith's study of the smaller Northern Region demonstrates how it is even more open to the effects of takeover and merger activities. ('The effect of external takeovers on manufacturing employment change in the Northern region between 1963 and 1973', Regional Studies, 13 (5), 1979, 421-38).

Doreen Massey has written of a 'new spatial division of labour' in U.K. manufacturing, between the predominantly white collar, control functions of multi-plant firms, concentrated in the South East, and their routine assembly functions, increasingly located in traditionally non-industrialised areas of the country where reserves of low-paid, mainly female labour can be tapped. This polarisation, she suggests, is taking place at the expense of the old-established industrial areas and conurbations, with their predominantly skilled manual, male, unionised labour forces (see D. Massey, 'In what sense a regional problem?', Regional Studies, 13 (2), 1979, 233-44). This hypothesis invokes many factors other than linkage to explain the movement of industrial investment away from old industrial areas, but the release of much modern production from dependence upon the high cost exchange of materials, and the growing ability of managers to communicate routine information over long distances have clearly permitted the new patterns of location to emerge. On the other hand, the importance of non-routine contacts between high-level managers, for the purposes of monitoring broad technological and market trends and long-term planning, imposes a positive stimulus towards the concentration of strategic functions into the South East, (see J.B. Goddard, 'The location of non-manufacturing activities within manufacturing industry' in F.E.I. Hamilton (Ed.) Contemporary Industrialisation, Longmans, 1978).

The geographical distribution of foreign ownership has also attracted attention, again because of its growth, and the assumption that it has significance for the ways in which local economies may be affected by remote decisions. So far, studies at the aggregate level have progressed little beyond descriptions of the distribution of directly foreign-owned manufacturing plants and their employment. Although there seem to be differences between the geographical patterns of American and European ownership, largely for historical reasons, the evidence for any distinctive form of spatial behaviour by overseas firms as a group, compared to U.K. based firms, remains unconvincing. (See H.D. Watts, 'The location of European direct investment in the United Kingdom' Tijdschrift voor Economische en Sociale Geografie, 71, 1980, 3-14; P. Dicken and P.E. Lloyd, 'Patterns and processes of change in the spatial distribution of foreign-controlled manufacturing employment in the United Kingdom, 1963 to 1975', Environment and Planning (A), 12, 1980, 1405-1426).

## 4. Linkage and Contact Patterns between Industrial Plants

The reasons for the significant changes which appear to be occurring in the aggregate location patterns of production and control in British manufacturing can thus only be inferred. They appear to be related to widely recognised changes in transportation costs, communications technology and the organisational structure of industry whose independent impacts are impossible to measure. Whether changing patterns of materials or information linkage are leading or following other developments cannot be determined directly at this scale. In any case, industrial linkage is most evidently important and measurable at the 'micro-' level of the individual establishment, where its influence on firms' behaviour can be judged in relation to other organisational factors. At this scale industrial linkage studies are nowadays concerned with patterns of individual plant organisation in relation to other locations, both within and outside the firms that own them. An associated problem is the impact of production changes at one location upon other plants, and on the community at large. The special significance to firms of local linkages, given much attention in earlier literature, is not particularly emphasized. In fact, although small-medium sized plants in specialist trades still tend to rely upon local contacts, patterns of supply and especially of sales linkage are today generally very widespread for most large establishments (M.J. Taylor and P.A. Wood, 'Industrial linkage and local agglomeration in the West Midlands metal industries' Transactions, Institute of British Geographers, 59, 1973, 314-36). Of course, this affects the nature and magnitude of the 'multiplier' on other local activities which may result from the establishment, growth or decline of plants in Assisted or other areas. A good deal of somewhat unnecessary disappointment has been expressed at the failure of imported investment to stimulate local industrial growth, even though local production linkages seem generally no longer to be a dominant feature of modern industrial activity. Local benefits mainly arise from the employment created in the processing of outside materials to produce widely marketed goods, and in associated service activities.

The literature on industrial linkage falls into two sections. The first, which has enjoyed a recent revival, is concerned with its general significance and measurement. An important paper by M.J. Taylor in 1975 summed up earlier discussions and has stimulated further debate more recently. In this, he suggested a model of linkage development and firm growth, associating the 'action spaces' of firms, defined by their current materials transactions, with their patterns of information linkage, and consequent investment ('Organisational growth, spatial interaction and location decision-making' Regional Studies, 9 (4), 313-24). Some of these ideas were subsequently developed in Taylor's studies of New Zealand industry. Another contribution, by A.G. Hoare, emphasized the notion that any measured pattern of linkages does not account for the potential variety of contacts that industrialists value in an area, nor their quality and reliability which have important psychological impacts upon business decision-making. ('Three problems for industrial linkage studies' Area, 10 (3), 1978, 217-21). Further complexities have been pointed out by R.T. Harrison, P.J. Bull and M. Hart, who criticize Taylor's model, especially for its assumptions about the processes by which industrialists obtain and act on information, and point out the logical problems of generalizing about different types of space (e.g. punctiform/continuous), reflecting different functions (e.g. routine/planning operations). Further debate on these issues is to be found in Area, 12 (2), 1980, 150-54.

Meanwhile, the second type of published work on linkages has been more concerned with empirical results, usually with reference to the implications of linkage patterns for regional industrial growth. Taylor augmented his earlier work on the West Midlands with a return survey of the region's foundry industry in 1976, providing a rare study of linkage change over an eight year period (M.J. Taylor, 'Linkage change and organisational growth: the case of the West Midlands Iron foundry industry' Economic Geography, 54, 1978, 314-36). The work of W.F. Lever in west central Scotland was particularly concerned with the multiplier effects of locally and externally owned plants in a cross section of industries (W.F. Lever, 'Manufacturing linkages and the search for suppliers and markets' in F.E.I. Hamilton (Ed.) Spatial Perspectives on Industrial Organisation and Decision Making, Wiley, 1974). McDermott also worked in Scotland, examining the relationship between large externally owned plants and indigenous enterprise in the electronics industry (P.J. McDermott, 'Ownership, organisation and regional dependence in the Scottish electronics industry', Regional Studies, 10 (3), 1976, 319-35). An early study of supply linkages to factories in west London was carried out in the mid 1960s by D.E. Keeble ('Local industrial linkage and manufacturing growth in outer London', Town Planning Review, 40, 1969, 163-88) but A.G. Hoare has more recently explored the wider implications of linkage patterns in the capital, including those generated by the dependence of firms upon Heathrow Airport ('Linkage flows, locational evaluation and industrial geography: a case study of Greater London' Environment and Planning (A), 7 (1), 1975, 41-58; 'Foreign firms and air transport: the geographical effect of Heathrow Airport' Regional Studies, (4), 1975, 349-68). Hoare has also published a study of linkages in overseas and locally orientated firms in Northern Ireland ('Industrial linkages and the dual economy: the case of Northern Ireland', Regional Studies, 12 (2), 1978, 167-80). An important study by J.N. Marshall has examined the effects of growing external ownership on manufacturing in the Northern region ('Ownership, organization and industrial linkage: a case study in the Northern Region of England', Regional Studies, 13 (6), 1979, 531-58). His conclusions suggest that although the regional pattern of materials transactions is relatively little affected by growing external ownership, information and business service contacts tend to be directed by large firms to their plants in other regions. Regional demand for consultancy and business services is thus reduced and in the long term, Marshall suspects, the quality of services available to locally based firms declines. This interrelationship between regional control, information sources and industrial competitiveness provides a more detailed insight into themes that have already been commented upon, from the aggregate studies discussed in Section 3.

Conclusion

Marshall's paper demonstrates a feature of recent work in British industrial geography that can be identified more generally; the conclusions from aggregate and 'micro-' level investigations of spatial behaviour and change are beginning to complement and reinforce each other. Linkage studies are increasingly being directed towards evaluating the working environments for manufacturing offered by various regions, through the availability of material, and especially of information contacts. The impact in other regions of the concentration of control functions into south east England, noted at the aggregate level, is being traced through micro-level studies. Unfortunately, of course, the relative magnitude and significance for employment of such changes are still unknown - they can be only hinted at by the evidence of broad patterns of ownership change, or by detailed studies of relatively small numbers of firms at the micro-scale. The same indirectness of evidence applies to the apparent correspondence between changes in the aggregate location patterns of manufacturing, reviewed in Section 1, and the types of firm behaviour described in Section 2. In spite of the important questions of evidence and interpretation which remain unanswered, industrial geography nevertheless offers a more or less coherent impression of the spatial and regional problems created by the decline of U.K. industry in the 1970s.

Much of the work has been orientated towards regional policy and, more recently, the decline of manufacturing in large cities. A number of new investigations are taking place into the latter problem and their results are likely to be published in the early 1980s. The problems of small, regionally based firms will also be given closer attention, as the result of a general interest in the local conditions which stimulate investment in established or new enterprise. Indigenous growth is likely to be more important for the economic success of regions in the future than the importation of jobs from elsewhere sought in the 1960s and early 1970s. Finally, future work is also likely to include more studies of location trends in the manufacture of particular products, especially in relation to technological changes and the actual local impacts of large company rationalisation programmes. The quality of employment offered by specialized production facilities in branch plants is now a matter of concern, as well as the numbers of jobs affected by such changes. Area-based studies will probably therefore emphasize more the effects of corporate and technical change in various industries upon the variety and types of skills required in local labour markets. Each of the four styles of work described in this chapter have made their contribution in the recent past towards our improved understanding of contemporary conditions in Britain, and each are likely to continue to do so in future.

SOURCES OF REGIONAL INFORMATION

Chapter Six

# AFRICA

Hazel M. Roberts

Just as Africa played a central role in the world of Ptolemy, Ibn Battuta, Henry the Navigator and Dr. Livingstone and his contemporaries, so today it is the focus of world attention in the struggle for economic and political development. It is the second largest of the continents, approximately 11,700,000 square miles, with about ten per cent of the population of the world (i.e. about 350 million) distributed over about one fifth of the world's total surface (remember that the amount of habitable land is reduced by such areas as the Sahara and Kalahari deserts). Oceans surround it - the Atlantic on the west, meeting the Indian Ocean in the south, the Indian Ocean and the Red Sea to the east, and in the north-east it is separated from the Sinai peninsula of Asia by the Suez Canal. The Mediterranean lies between Europe and Africa.

The highest point is Mount Kilimanjaro, 19,340 feet, and the lowest is Lake Assal (in Djibouti), which is 509 feet below sea level. It has fewer high mountains and fewer low-land plains than any other continent. Most of the land below 500 feet is within 500 miles of the coast. The Great Rift Valley, which extends about 4,000 miles, is the most distinctive relief feature. Practically the whole of the continent lies within the tropical zone, with the Equator bisecting it, though as it is wider in the north the influence of the sea in southern Africa extends further inland.

Man has had a disastrous effect on both vegetation and animal life, with increasing population and the provision of better tools, including guns. Many countries have set aside large areas as national parks, game or forest reserves, especially in East and Central Africa.

There are many countries, differing greatly in terms of development, urbanisation and literacy, with a variety of tribes and languages, even within one country - often causing problems of internal unity. Africa north of the Sahara is inhabited mainly by people speaking Semitic and Hamitic languages, and the Bantu-speaking people extend over most of the continent south of the Equator. The number of separate and distinct languages has been estimated from 800 to more than 1,000. Self-government and political independence from colonial powers have been achieved in most countries, but there is still a struggle for freedom of the indigenous peoples in the south. The strongest power in Africa, Nigeria, has recently drawn attention to what it calls the centre piece of its foreign policy - 'the total liberation of all colonised parts of Africa' - hopefully by a mixture of economic and diplomatic pressure, rather than war.

Africa has been explored from the time of the Phoenicians (who founded the colony of Carthage on the North African coast about 820 B.C.), by the Greeks and Romans (the Greeks founded Cyrene about 630 B.C.), the Arabs and the Portuguese. By the sixteenth century the English were venturing round the west coast of Africa. The stories of British explorers to the interior, e.g. Mungo Park, Burton and Speke, mainly searching for the sources of some of the world's greatest rivers, have fired the imagination of every reader. David Livingstone changed his interest from missionary work to exploration, believing that opening up the interior was the way to destroy the slave trade and break Africa's isolation from the rest of the world. Eventually came the scramble for Africa from various European powers, with colonisation and settlement following.

## The study of African geography

Many books dealing with African geography begin with a section on the background to Africa's exploration and discovery. B.E. Thomas, writing about the geographical exploration of tropical Africa during the nineteenth century, in *The African World, a survey of social research*, ed. R.A. Lystad (A.S.A., Pall Mall Press, 1965), suggests that there was a transition from geographic exploration to geographic fieldwork or field studies - 'the exploration period for Africa was about over by the first decade of this century....travels by explorers have given way to scientific investigations by geologists, geographers, biologists and anthropologists' (p.249). A standard bibliographic work, which has not been superseded, is E.G. Cox's *Reference Guide to the Literature of Travel, including voyages, geographical descriptions, adventures, shipwrecks and expeditions*. (3 vols., Univ.of Washington Press, 1935-49). An account of different people's travels and 'mercantile expeditions' can be found in *A Geographical Survey of Africa* by J. M'Queen (1st edn 1840, new impression Frank Cass, 1969).

When people explored they made maps, and different techniques of mapping followed, from the primitive maps of the African coast made during the fifteenth and sixteenth century discoveries, to seventeenth century cartography with maps for navigation, surveying, and physical geography. R.V. Tooley's *Collectors' Guide to Maps of the African Continent and South-*

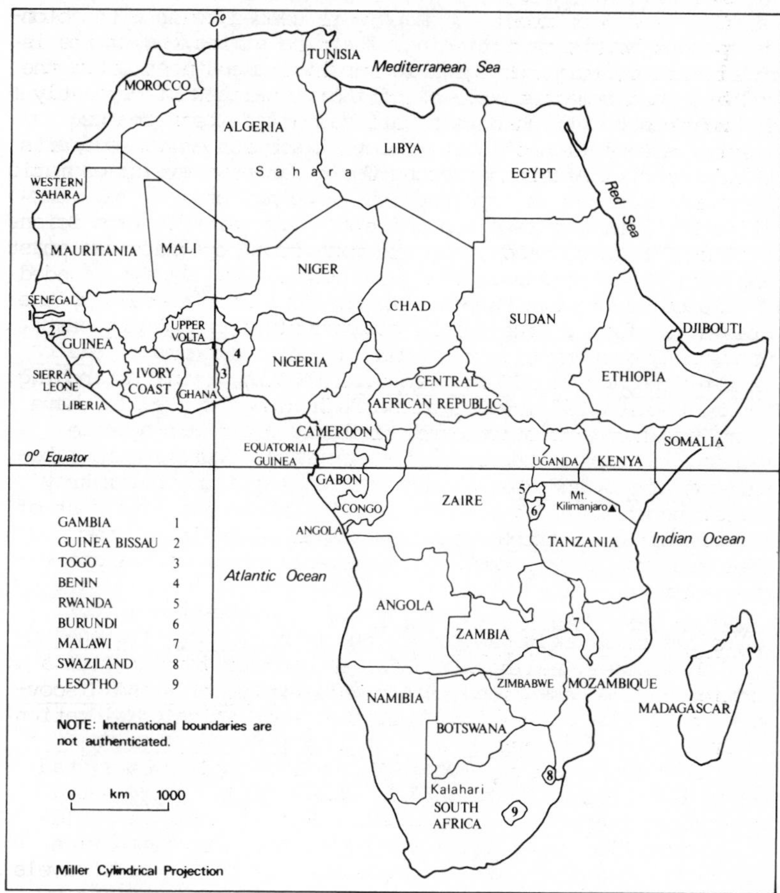


**Figure 2. AFRICA**

ern Africa (Carta Press, 1969) is an annotated list of early printed maps - up to 1886. Maps were needed to show the boundaries during the partitioning of Africa at the end of the nineteenth century, topographic maps followed, then general atlases, and detailed atlases of individual countries. Only a few African countries have a national atlas, but some general atlases are mentioned in the relevant section at the end of this chapter. After the Second World War new developments included progress on topographical mapping, and now maps for different kinds of research and uses, e.g. distribution of railways, houses, roads, etc. are made from topographical maps and aerial photographs.

Studies of what happened in the past help explain the present surface of the earth, both physical changes by natural processes, and the effect of man on the environment. Africa in the Iron Age: c 500 B.C. to A.D. 1400 by R. Oliver and B.M. Fagan (C.U.P., 1975) makes a thorough study of the African past, and has suggestions for further reading arranged by region. A useful historical background is found in the first part of Africa and the Islands by R.J. Harrison Church (4th edn, Longmans, 1977).

Physical geography includes the study of landforms, weather and climate systems, soils, vegetation and water resources. Although it was written particularly about Asia, The Physical Geography of the Tropics by J.G. Lockwood (O.U.P., 1976) has much useful material, and Geomorphology in Deserts by R.V. Cooke and A. Warren (Batsford, 1973) investigates the nature of landforms, soils and geomorphological processes in deserts. Colin Buckle writes specifically on geomorphology in Africa in Landforms in Africa: an introduction to geomorphology (Longmans, 1978). Possibly the best chapters on the physical environment are in J.I. Clarke et al, An Advanced Geography of Africa (Hulton, 1975).

Population and settlement geography is concerned with the distribution and movement of population, and the relationship of people to the physical and economic environment through employment and food production. Research in this area includes mapping of population distribution, collection and analysis of census data and the determining of settlement patterns. Two books linking populations with the physical environment are Human Ecology in Savanna Environments, ed. David R. Harris (London Academic Press, 1980) and Man's Environmental Predicament, an introduction to human ecology in Tropical Africa by Denis F. Owen (O.U.P., 1973). R.P. Moss and R.J.A.R. Rathbone edited The Population Factor in African Studies (Univ.of London Press, 1975), the proceedings of a conference organised by ASAUK. Another book on the subject is Populations of the Middle East and North Africa by J.I. Clarke and W.B.Fisher (Univ. of London Press, 1972) and Clarke also wrote Population Geography and the Developing Countries (Pergamon Press, 1971). Latest population statistics are available in the

Demographic Yearbook of the United Nations. Some bibliographies on settlement are listed in the bibliographic section in this chapter.

Medical geography is the study of disease and its causes, and its relationship to the environment. Information is provided about the natural environment and the mapping of diseases, in one area, by B.W. Langlands' article, 'Maps and Medicine in East Africa' in Map Librarianship: Readings comp. by Roman Drazniowsky (Scarecrow Press, 1975) and also in Special Libraries Association, Geog. and Maps Division, Bulletin No.78 (1969), 9-15. There is a chapter in L.D. Stamp's Africa: a Study in Tropical Development (3rd edn, Wiley, 1972) called 'The plagues of Africa - Pests and Diseases'. A study of one particular disease and its cause is J. Ford's The Role of the Trypanosomiases in African Ecology: a study of the tsetse fly problem (Clarendon Press, 1971). Another study, about malaria, is the book by R. Mansell Prothero, Migrants and Malaria (Longmans, 1965). There is a chapter on 'Human Diseases' in the book by D.F. Owen Man's Environmental Predicament (O.U.P., 1973).

The previous two sections look at human or social aspects of geography, whereas economic geography includes agriculture, minerals, transport, trade and industry. Agriculture is closely related to physical conditions (i.e. temperature, rainfall and soil) and social features like eating habits, and farming techniques. A background to land-use including a historical outline, is found in L.D. Stamp (ed.), The History of Land Use in Arid Regions (UNESCO, Paris, 1961). The study of agriculture can be made by commodity, or by regional investigations, and so for example can the study of mining - which is dependent on the physical conditions of geological structure as well as availability of labour, both skilled and unskilled. K.R.M. Anthony's Agricultural Change in Tropical Africa (Cornell U.P., 1979) is a basic text, and there is the earlier J.C. De Wilde's Experiences with Agricultural Development in Tropical Africa (Johns Hopkins Press, 1967 - 2 vols). Various FAO (Food and Agriculture Organisation, United Nations) publications give specific information on economic studies (as well as many other matters), e.g. FAO World Conference on Credit for Farmers - series Credit Markets of Africa: Agricultural Credit for Development (Rome, 1975). Economists emphasise economic theory, market conditions and financial aspects, whereas geographers look more at natural and cultural conditions and resources, regional links and differences, patterns of production and trade. The reader should turn to B.W. Hodder's Economic Development in the Tropics (3rd edn, Methuen, 1981). A book which gives information on environmental, economic and political problems is Africa Today: a short introduction to African affairs, also by B.W. Hodder (Methuen, 1978); and A.M. O'Connor's The Geography of Tropical African Development: a study of spatial patterns and economic change since independence (2nd edn, Pergamon, 1978) looks at the economics of topics such as mining and agriculture.

The exploration of Africa, and its colonisation and partition, have been closely linked with trade and transportation. The problems of tropical climates, disease, lack of harbours and deltas, have all been important factors in this aspect of geography. Attention has always been given to problems of developing roads, railways and harbours, from colonial days and by the independent nations, see for example the chapter on 'Transportation in Africa' in L.D. Stamp, Africa: a Study in Tropical Development (3rd edn., Wiley, 1972). B.E. Thomas in R.A. Lystad's The African World (A.S.A., 1965) says that to answer the question of why transportation geography occurs, it is necessary to look at the means of transportation - animals or vehicles - i.e. how and why; the nature of goods carried - i.e. what and why; and the routes followed and the resulting patterns on the earth's surface - i.e. where and why. Many geographical aspects are linked in such studies - climate, terrain, culture, commerce, mapping and statistics.

Urban geography includes the study of settlement forms and patterns, e.g. the size and distribution of cities, their situation, functions, and relations to other areas. A.L. Mabogunje's Urbanization in Nigeria (Univ. of London Press, 1968) studies trends and patterns,and the book, The City of Ibadan ed. by P.C. Lloyd, A.L. Mabogunje and B. Awe (C.U.P., 1967) looks at the life, the work and the people of this city in Nigeria, and includes a glance into future problems. W.A. Hance's Population, Migration and Urbanization in Africa (Columbia U.P., 1970), though less well known than his standard work on Africa, The Geography of Modern Africa (2nd rev. edn., Columbia U.P., 1975) (mentioned later), repays study. There is now a bibliography on urbanization, by A.M. O'Connor (mentioned in the section on bibliographies) and he is currently writing The City in Tropical Africa, a geographical study to be submitted to Hutchinsons, for publication in the near future.

Literature on political geography is increasing because of interest in human geography, in political affairs, and in Africa generally. Early studies looked at the origins of African boundaries and how they encompassed new political areas, changing the patterns of population and trade, and the effect of dividing tribes, tribal land units, forests and game reserves on boundaries. Later aspects of boundaries on which political geographers work are delimitation, demarcation, and the advantages or problems of different types of boundaries like rivers (which vary seasonally and change course) or mountain ranges. The basic reference is E.A. Boateng's A Political Geography of Africa (C.U.P., 1978), and his earlier book, Independence and Nation Building in Africa (Ghana Publishing Corp., 1973), traces the course of the movement towards independence since the Second World War. B.W. Hodder's Africa Today (Methuen, 1978) also looks at political problems. Periodicals are important sources of information for articles on current developments, e.g. the bi-monthly journal Commonwealth recently included an article on 'The new Nigeria - political background to the new leadership', and the monthly New Internationalist looks at the politics of poverty and development.

'In geography, it is sometimes said that the systematic fields are sciences of the order of the natural or social types, but that regional geography is an art.' (Thomas in Lystad, p.27).

There are many approaches to systematic geography, and a flow of knowledge between other fields and disciplines. To study Africa using the regional approach would mean taking one region first, studying all the elements of geography, physical and human, within that region. The boundaries would either be political units (on the grounds that people are familiar with them), or divisions by characteristics of the earth's surface either by physical feature or man's influence, i.e. tribal cultures or land use patterns. This could include large regions like the Sahara, the Eastern Highlands, West Africa, or small regions, e.g. 'the Souf' and 'the Mzab' in the Sahara. Studies like this mean that geographers may have to devote years of field work and research to a particular people to become experts on that area, and therefore this is a time-consuming way of producing information. However, there is still demand for regional geographic courses on Africa, and for information on regions, as provided in Part 2 of L.D. Stamp's Africa: a Study in Tropical Development (3rd edn, Wiley, 1972), or R.J. Harrison Church's Africa and the Islands (4th edn, Longmans, 1977). Many regional studies select the most important topics, or those already surveyed, e.g. agriculture, economics and urban studies. Other regional studies use systematic methods for studying a country, e.g. climate, landforms, vegetation and land use. Another way is to treat physical features systematically, and human features regionally. The section in this chapter on books discusses these and other books further.

Most geographical research on Africa in the early part of this century came from Britain, France, Germany and Italy (the colonial powers), beginning with work on physical geography, and then including economic and social geography. French geographers produced regional geographies on the French colonies of Africa from as early as the last century. American research has centred mostly on economic aspects. Many more books are appearing now based on field work, on such subjects as culture, agriculture and industries. Colleges and universities were established in tropical Africa with geography departments and research institutes providing facilities for geographical studies. World attention has been focussed on the economic, political and regional geography of Africa with increase in African trade, and emphasis on economic planning and development with the independent nations. P. Gould in the chapter 'Geography, Spatial Planning and Africa: the responsibilities of the next 20 years' in Expanding Horizons in African Studies by G.M. Carter and A. Puden (Northwestern U.P., 1968), writes about the importance of research, and the lack of feedback from academic research to governments and agencies of countries where research is undertaken; provision of information and sound theory are major geographical tasks in Africa.

The use of information tools

This section looks at different types of material which provide information on African geography, or are written specifically about aspects of geography in Africa; where this material can be located; and where it is studied. Some parts of books date quickly, and it is important to keep up with recent work because of constant changes, especially for example in economic and political geography. Several authors mention the lack of literature on the basic geography of Africa, or the fact that sources of information are scattered, but most provide bibliographies with their books or chapters as a guide for the reader to further research.

The Directory of African Studies in U.K. Universities comp. by R.E. Bradbury (ASA UK, 1969) is a starting point for places where geography of Africa is studied. It was compiled at the request of the African Studies Association, with information supplied from a questionnaire. It is a general guide to courses and programmes available in about 40 universities and institutes, of which about 30 teach geography, and has an index to disciplines represented in staff lists. There is also a list of non-university institutes concerned with Africa. The information was gathered between 1967 and 1969. The hoped for new editions have not appeared but there are updates in the journal African Research and Documentation, No. 18 (1978), No.19 (1979), and some further additions in No.20 (1980). There is also a small booklet produced by SOAS (School of Oriental and African Studies, University of London) called African Studies in London (1976), and the Geography Department in SOAS produces and regularly updates a 'Programme for the B.A. in Geography' which gives course outlines and basic reading lists. The courses include geography of Africa south of the Sahara, urban geography with special reference to Asia and Africa, agricultural development problems, concepts and techniques of development planning. A similar A.S.A. publication for the U.S.A. is a Directory of African Studies in the U.S. which is published twice yearly. The 6th edition (1979) comp. by D. Duffy and M.M. Frey, includes about 900 universities and colleges offering courses on Afro-American and African studies. SCOLMA (Standing Conference on Library Materials on Africa, U.K.) lists institutes and African Studies Centres in U.K., America, and throughout the world in Directory of Libraries and Special Collections on Africa, comp. by R. Collison and revised by J. Roe (3rd edn, Crosby Lockwood, 1970).

Periodicals include publications issued in successive parts, and besides journals, newspapers, government publications and bulletins are worth consulting for current affairs and developments. A great deal of material, or information on current research, is published first in periodicals. Articles on African geography appear from time to time in a vast number of journals. Sanford H. Bederman in Africa: a bibliography of geography and related subjects (3rd edn, Georgia State Univ., 1974) has a comprehensive list of periodicals consulted, including many from African countries, with a very useful index of topics by country. J. Asamani's Index Africanus (Stanford, Hoover Institution, 1975) lists over 200 journals concerned with Africa and 23,000 articles published 1885-1965. The Library of Congress, African Section, produced Africa South of the Sahara: index to periodical literature 1900 - 1970 (Washington, 1971 + Suppl. 1973) with about 100,000 items. British journals include the Geographical Journal, Geography, and Progress in Human Geography which recently contained an article on 'The Environmental Factor in African Studies' by Paul Richards (Vol.4, No.4, 1980). Popular articles with many illustrations appear in the monthly Geographical Magazine. In America are published Economic Geography and the Geographical Review. The issues of Focus, each devoted to a single country or region and published by the American Geographical Society, monthly, are handy sources of up-to-date information, e.g. two issues in 1977 carried parts 1 & 2 of 'Ivory Coast: political stability and economic growth'. Other sources are Cahiers d'Etudes Africaines (Paris), and Cahiers d'Outre-mer (Bordeaux). Non-geographical quarterlies which occasionally have articles of interest to geographers are African Affairs and the Journal of Modern African Studies. The previously mentioned New Internationalist deals with 'the people, the ideas and the action in the fight for development', and reports on issues of world poverty, while Commonwealth is another source for current affairs articles.

Many books appeared in the 1960's (several with later editions in the 1970's), all contributing to the knowledge of the geography of Africa and geographical aspects of Africa's problems. This constituted what R.M. Prothero called a 'second scramble for Africa' headed by the publishers, and was in response to the growth of knowledge about Africa and the demand for that information. British writers tend to follow more traditionally established geographical practice - discussing systematic elements first - i.e. climate, vegetation, landforms, and then looking at various parts of Africa, usually on the basis of national divisions. Most authors begin with an explanation and justification for tackling Africa in their particular way - which does tell the student the book's emphasis. The best of the systematically organised texts are J.I.

Clarke's An Advanced Geography of Africa (Hulton, 1975), and C.G. Knight and J. Newman (eds), Contemporary Africa (Prentice-Hall, 1976). L.D. Stamp and W.T.W. Morgan in Africa: a Study in Tropical Development (3rd edn, Wiley, 1972) look at the African continent under systematic headings (some previously mentioned in this chapter),and then in part 2 at the countries and regions of Africa. Another book with a similar approach is Africa and the Islands by R.J. Harrison Church (4th edn, Longmans, 1977), with regional studies of areas like North West Africa, North East, West, Central, Eastern and Southern Africa, discussing political, social and economic characteristics and trends in the regions. The author states that a book on Africa needs to be a joint effort by specialists in certain areas and aspects. The purposes of W.A. Hance's book, The Geography of Modern Africa (2nd rev. edn, Columbia U.P., 1975), is to look at the economy of Africa and see what affects economic development, and to assess potential for growth. He uses the regional approach, with chapters devoted to individual countries or groups of countries. H.J. De Blij calls his book a regional text on Africa - A Geography of Subsaharan Africa (Rand McNally, 1964) - but it is different from others in that it takes a thematic approach, using the political boundaries as dividing lines, looking at a specific aspect of a particular country rather than general geographic qualities, e.g. land in Kenya, political geography of federation in Nigeria. It is now superseded by African Survey, A.C.G. Best and H.J. De Blij (Wiley, 1977). A.T. Grove in Africa (3rd edn, O.U.P., 1978)- published originally as Africa South of the Sahara in 1967 - is concerned with the old and the new side by side, writing a geography of modern Africa taking into account the traditional and historical background. R.M. Prothero, editor of A Geography of Africa: regional essays on fundamental characteristics, issues and problems (Routledge & Kegan Paul, 1973 reprint with rev. final chapter), tries to break away from the established textbook with systematic approach followed by reional divisions, and produces a series of regional essays on areas varying considerably in size and population. The authors have developed their essays by considering a theme or themes most relevant to that region, aiming to combine contemporary and fundamental issues. The book looks at ideas rather than facts under traditional headings, e.g. 'Southern Africa: Bonds and Barriers in a Multi-Racial Region', 'West Africa: Growth and Change in Trade'. B.W. Hodder and D.R. Harris in Africa in Transition (Methuen, 1967) bring together essays by several authors on regions of Africa, with the historical and physical setting, and development up to the time of writing. There is a gradual increase of writings by people within Africa, e.g. J.S. Oguntoyinbo et al, A Geography of Nigerian Development (Heinemann, 1980), and Regional Planning and National Development in Tropical Africa (Ibadan U.P., 1977) ed. by A.L. Mabogunje and A. Faniran; and perhaps one of the most notable recent geographic studies of South Africa is A. Lemon's Apartheid: a geography of separation (Saxon House, 1977). Although this chapter focusses largely on tropical Africa, and on English language material, there is also a vast literature in French.

Bibliographies - One especially for geographers is S.H. Bederman's Africa: a bibliography of geography and related subjects (3rd edn, Georgia State Univ., 1974), (formerly A Bibliographic Aid to the Study of the Geography of Africa) which arranges books and articles alphabetically by author in region, and also has a topic index for each country. General bibliographies on Africa include the International African Bibliography (Mansell, 1971 - ) a quarterly produced at SOAS, but previously a section in the International African Institutes's journal Africa, 1929-1970. Material is arranged geographically by country, and there is also a subject tracing at the end of each entry. Current Bibliography on African Affairs (Washington, African Bibliographic Center, 1962 - ) is also a quarterly and useful as a quick current awareness tool. H.F. Conover's Africa South of the Sahara: a selected annotated list of writings (Library of Congress, Washington, 1963) and K.M. Glazier's Africa South of the Sahara, a select and annotated bibliography, 1958-1963 (Hoover Institution Bib. Series, 16, 1964) are useful for background information up to the time of their publication.

Specialized bibliographies include A.M. O'Connor's Urbanization in Tropical Africa: an annotated bibliography (G.K. Hall, 1981) covering French and English publications from 1960 on, arranged by area, and with about 2,500 items. Another publication in G.K. Hall's series of bibliographies and guides in African studies is Farming Systems in Africa: a working bibliography, 1930-1978 (1979) by S.M. Lawani, F.M. Alluri and E.N. Adimorah; and the African Bibliographic Center is also producing a series, one of which is A Bibliography of African Ecology comp. by D.J. Rogers (No.6, Greenwood Press, 1979). FAO (U.N.) has produced a Bibliography on Land Settlement (Rome, 1976), and there is a recent series (1980) of Draft Bibliographies on Human Settlement by the Univ. of Lund (Sweden) School of Architecture, Dept. of Building Function Analysis. The 4 volumes so far are 'Developing countries with emphasis on East Africa', 'Kenya', 'Mozambique', and 'Zambia'.

Two further general bibliographies from which it is possible to extract reference to articles on aspects of geography are Hans Panofsky's A Bibliography of Africana (Greenwood Press, 1975) which has a guide to resources by subject and discipline with relevant sections under Sciences (Agriculture) and Social Sciences (Demography, Urbanization, Economics), and The African Experience ed. by John N. Paden and Edward W. Soja (3 vols., Northwestern U.P., 1970).

Theses - The A.S.A. of the U.S.A. produces a Research in Progress: a selected list of current research on Africa, annually, from 1970/71 on, and M. Bratton's American Doctoral Dissertations on Africa, 1886-1972. The equivalent U.K. publication is Theses on Africa, 1963-1975, accepted by Universities in the U.K. & Ireland, comp. by J.H. St.J.McIlwaine for SCOLMA (Mansell, 1978). SCOLMA also produced an earlier listing of U.K. theses. University Microfilms (Ann Arbor) lists all American theses and dissertations and publishes topic indexes.

Other information sources include handbooks, encyclopaedias, dictionaries and yearbooks. A book which calls itself a handbook, and is also a useful bibliography, is Margaret Killingray's African Studies: a Handbook for Teachers (SOAS, Extramural Division, 1979). It has a section on Geography and Integrated Studies, and outlines topics for teaching as an alternative to the teaching of facts region by region; a bibliography of audio-visual materials is included. A similar publication from the UCLA African Studies Center is Teacher's Resource Handbook for African Studies: an annotated bibliography of curriculum materials by J.N. Hawkins and J. Maksik (UCLA Occas.Paper No.16, 1976). The last section provides information for the older school student. Handbooks or surveys are concerned principally with the situation in a given area at the time of publication, e.g. Black Africa: a comparative handbook (Free Press, Macmillan, 1972) by D.G. Morrison et al. There are comparative and country profiles - under headings such as Urban Patterns, Political Patterns - and maps of each country. Encyclopaedias are general reference works and are useful for outlines of themes or regions. Besides a large section on Africa in the New Encyclopaedia Britannica (15th edn, 1974), there is an 11 volume Standard Encyclopaedia of Southern Africa (1970-1975), and MacDonald's Encyclopaedia of Africa - secondary school level (MacDonald Educational, 1976) which is organised in thematic sections.

Dictionaries are useful for definition of terms, e.g. there is a series of African Historical Dictionaries (Scarecrow Press, 1974 - ), with now more than 20 completed for different countries, which include people and places and contain extensive bibliographies. The Europa Yearbook, Africa South of the Sahara (1971 - ) contains more general historical material than most, plus current factual data on each country, and articles on background to the continent, e.g. 'Political and Social Problems of Development', 'Agriculture in African Economic Development'. There are sections on physical and social geography for every country. The standard work for recent developments is Africa Contemporary Record, annual survey and documents by Colin Legum and John Drysdale (Africa Research Ltd, 1968 - ). J. Harvey's Statistics - Africa: sources for market research (CBD Research, 1970 - ) is an annual mainly intended for business use.

Atlases - Philips' Modern College Atlas for Africa ed. by H. Fullard (12th edn, G. Philip, 1974) is a general atlas with emphasis on Africa, and another world atlas with emphasis on southern Africa is Senior Atlas for Southern Africa (Collins-Longman, 1976). The best atlas is the Atlas of Africa, Editions Jeune Afrique, ed. by Van Chi-Bonnardel (Paris, 1973), with both French and English editions. It contains two maps of each country showing physical features, administrative divisions, the economy, and is accompanied by text. A most valuable recent addition to the atlases of Africa is the Cultural Atlas of Africa (Phaidon, 1981), compiled by Jocelyn Murray. This is a work that should certainly be in any library of African studies and is a suitable colophon to this short introduction to the literature on the continent.

Chapter Seven

# SOUTH ASIA

W.J. Alspaugh

The subcontinent of South Asia consists of the nations of India, Pakistan, Bangladesh, Nepal, Bhutan and the Maldives. To these we may add parts of Afghanistan, Tibet and Burma that clearly fall into the Indic civilizational sphere. Within the scope used in this essay, South Asia can be considered a unified world region on geographic, historical and socio-cultural grounds. This region is the home of one-fifth of the world's people and the site of one of its oldest continuous civilizations. Although its conflicts and crises have in the past been largely of regional significance only, the subcontinent is rapidly being drawn into the mainstream of world politics as both the Indian Ocean and the mountain frontiers become objects of great power rivalry.

This essay will focus on publications produced in South Asia and on Western sources which are primarily or exclusively concerned with the subcontinent. It should be noted that much of the best scholarship on South Asia is published in the West and is accessible through the standard bibliographic tools, which offer only very limited coverage of South Asian publications from South Asia itself. The quality of South Asian publications varies enormously in materials, printing and scholarship. Many "scholarly" works lack footnotes, indexes and bibliographies, and many reference works are poorly produced A-to-Z listings without classification. Nonetheless, there is a considerable amount of first-rate scholarship and much more that is of use to the researcher.

South Asia is one of the great world centres of English writing and publication, India alone being third after the U.S. and U.K. English continues to be an official language of national and state governments, the major medium of higher education, and a "link language" between the many linguistic regions. Thus documents, reports, and social science literature in English are abundant. However, anyone aspiring to be a South Asia specialist will require a knowledge of one or more vernaculars, especially if researching regional or local problems.

India dominates the subcontinent bibliographically, as well as politically and economically. Not only is its volume of publications much greater than that of the other nations, but India alone has a well-developed and extensive bibliographic apparatus. The researcher interested in the smaller nations will certainly want to consult the Indian sources, while the reverse is usually not the case.

## Growth of South Asian Studies

After the era of European travellers' tales of the marvels of the East, serious Western study of South Asia began in the years before 1800 with the philological and philosophical investigation of classic texts - the kind of scholarship commonly known as orientalism or indology. Later in the nineteenth century came empirical studies of Indian and Ceylonese society, at first by officials and missionaries, later in the twentieth century by a few Western social scientists who ventured to India for fieldwork, or by their native students. In series such as the Ethnographic survey, Archaeological survey, and Linguistic survey of India, sponsored and published by the Government of India, these studies soon outgrew any pretence of relation to administration, establishing a continuing tradition of public support of scholarship.

The end of World War II and the independence of the former colonies were followed by a massive expansion of Asian studies. In the United States and the United Kingdom, this growth was fuelled by generous financial inputs from governments and private foundations, in South Asia by the overall growth of higher education, by nationalist sentiment, and by the prospect of using "applied social science" in attacking national problems. During this period, many universities throughout the world set up programmes on South Asia, or more general programmes in Asian, or Oriental, studies. By the late 1960s, the expansion appeared to have ended. With diminishing outside support, especially the end of the NDEA and PL-480 programmes in the U.S., many schools are faced with the problem of retrenchment. However, Western academic resources expended on South Asia are still far less than those devoted to East Asia, and although programmes may have exceeded the short-term demand for Ph.D's, South Asia studies must continue to expand if they are to be commensurate with the importance of this area in the world.

## Library Guides and Catalogues

There are very few libraries in Europe and America which maintain extensive collections on South Asia. In 1978, of twenty-eight U.S. libraries participating in the U.S. Library of Congress Foreign Acquisitions Program ("PL-480) for South Asia, only eighteen acquired Indian English publications on a comprehensive basis, while only four American libraries maintained comprehensive levels in all languages. The scholar

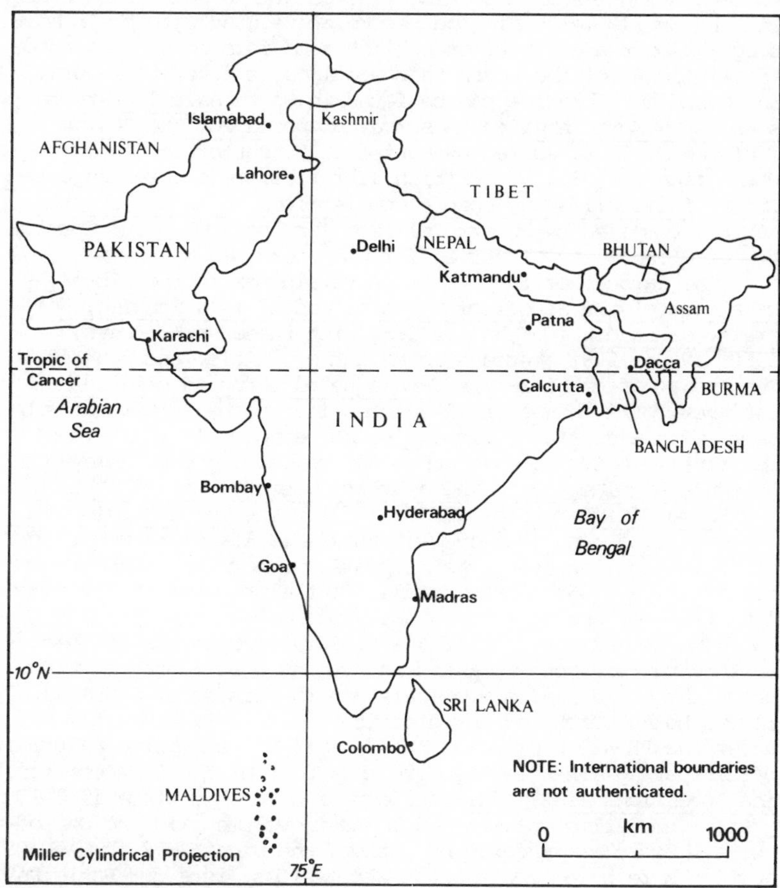


Figure 3. SOUTH ASIA

contemplating research should first determine what collections will be available to him and become acquainted with their holdings; almost all libraries publish some form of written guide. Descriptions of the South Asia holdings of forty-three U.S. and Canadian libraries can be found in South Asia library resources in North America; A survey prepared for the Boston Conference, 1974, edited by Maureen L.P. Patterson (Inter Documentation Co., 1975). British libraries are covered by Directory of libraries and special collections on Asia and North Africa, compiled by Robert Collison for the SCONUL Sub-Committee of Orientalist Libraries (Archon Books, 1970).

The largest of the published catalogues is the Library catalogue of the School of Oriental and African Studies, University of London (G.K.Hall, 1963, with three supplements, 1968, 1973, 1979). Access is by author, title, and subject. The India Office Library's Catalogue of European printed books includes works in Western languages indexed by author and subject, with a separate volume listing periodicals. A valuable descriptive guide to this important collection is Stanley C. Sutton's A guide to the India Office Library, with a note on the India Office records (HMSO, 1971). The Union Catalogue of Asian publications 1965-1970 (Mansell, 1971) is an author list of books acquired by British libraries other than SOAS. South Asianworks are 46% of the total. Of the promised annual supplements, only one, for 1971, has appeared.

The Accessions list of the Library of Congress has become a standard reference work and is available at libraries throughout the world. Also available are the lists for South Asia record books purchased for distribution to U.S. libraries under the PL-480 programme, but it must be remembered that most participating libraries receive only part of the LC accessions. The Accessions list: India has appeared monthly since 1962 with yearly cumulative volumes and separate volumes listing periodicals. Books are indexed by author, and since 1972, by subject. There have been separate Accessions lists for Pakistan, Bangladesh, Sri Lanka and Nepal, but after January, 1981 all have been consolidated into a single Accessions list: South Asia.

The Catalog of the Ames Library of South Asia, University of Minnesota (G.K. Hall, 1980) has just been published, and the publisher has indicated interest in publishing other South Asian collections. Substantial holdings related to the subcontinent can be found in the New York Public Library's Dictionary catalog of the Oriental collection (G.K. Hall, 1960, with supplement, 1976). The South Asia Microfilm Project is maintained by the Center for Research Libraries in Chicago to make available older, hard-to-obtain monographs and periodicals, as well as government publications such as land settlement reports, and archival materials. The latest edition of its SAMP catalog (CRL, 1980) lists 2096 items, with author and subject index. Substantial sections on South Asia are found in the regional volumes of the Research catalogue of the American Geographical Society (G.K. Hall, 1968, with supplements, 1972, 1978).

There are no up-to-date library catalogues from South Asia itself. The National Library in Calcutta has published its Author catalogue of printed books in European languages (the Library, 1964).

## Bibliographies of Reference Works

The most valuable of these tools is A guide to reference materials on India, by N.N. Gidwani and K. Navalani (Jaipur, Saraswati Publications, 1974). In spite of its name, this two-volume annotated guide cites sources on all South Asian nations, including bibliographies from monographs and journals as well as those in book form. Two short but very useful recent works are Kanta Bhatia's Reference sources on South Asia (South Asian Regional Studies, U. of Pennsylvania, 1978), and Iqbal Wagle's Reference aids to South Asia (University of Toronto Library, 1977). For India there is A bibliography of bibliographies on India, by D.R. Kalia and M.K. Jain (Concept, 1975), and Indian reference sources; an annotated guide to Indian reference books, by H.D. Sharma, S.P. Mukherji, and L.M.P. Singh (Indian Bibliographic Center, 1972).

## General Retrospective Bibliography

The most current and complete single-volume bibliography is South Asian civilizations: a bibliographic synthesis, edited by Maureen L.P. Patterson (U. of Chicago Press, 1981). Its 28,017 entries - books, articles, dissertations - are arranged in a detailed topical outline based on eight geo-cultural areas. Only general, non-technical works on geography are cited, but many related books can be found in the social sciences sections.

An impressive and useful body of bibliographic sources, now totalling nineteen volumes, is sponsored by the Indian Council of Social Science Research. Published by various firms, and not formally indentified as a series, each of these volumes is titled A survey of research in..; thus far, volumes have appeared for sociology and social anthropology (3v.), geography (2v.), economics (6v.), public administration (2v.), psychology, management (2v.), demography, and political science. Each book consists of a series of "trend reports" on various sub-areas, with appended bibliographies, each report prepared by a recognized authority.

South Asia bibliography: a handbook and guide, edited by J.D. Pearson (Harvester Press, 1979) is a collection of essays, covering different types of materials, subject areas, and countries. Jean-Luc Chambard's Bibliographie de civilisation de l'Inde contemporaine is strongest in the areas of recent social and economic change, but its usefulness is impaired by the lack of indexes

For the Islamic nation states, Pakistan and Bangladesh; Bibliographic essays in social science, edited by W. Eric Gustafson (U. of Islamabad Press, 1976) is the most up-to-date source. Uta Ahmed's Bibliographie des Deutschen Pakistans-Schrifttums (Deutsch-Pakistanisches Forum, 1975) includes English publications from the subcontinent as well as those in German and by German writers. Satyaprakash has compiled two bibliographies for the Indian Documentation Service: Pakistan; a bibliography 1962-1974 (IDS, 1975) and Bangla Desh; a select bibliography (IDS, 1976); both are limited to books and articles published in India.

For Sri Lanka the standard source is H.A.I. Goonetileke's excellent A bibliography of Ceylon (Inter Documentation Co., 2 vols., 1970, with supplementary v.3, 1973). Equally excellent is the Bibliographie du Népal, edited by L. Boulnois, and others (Centre national de la recherche scientifique, 1969): volume 1, Sciences humaines, has a supplement covering 1967-73 (published 1975); volume 3, part 1 indexes maps of Nepal; volume 3, part 2 covers botany; volume 2 has not yet been published. Two other useful sources on the Himalayan kingdom are the Bibliography of Nepal, edited by Khadga Man Malla (Royal Nepal Academy, 1975) and A bibliography of Nepal, edited by Basil C. Hendrick, and others (Scarecrow Press, 1973).

There are a number of bibliographies covering states, provinces, and regions within the subcontinent. These can be found in the guides to reference works above. In general, they emphasize history, ethnology and economics, with little coverage of geography *per se*.

## Current Bibliography

One of the best and most up-to-date sources is the Guide to Indian periodical literature, published since 1964 by the Indian Documentation Service. Appearing quarterly with annual accumulations, it indexes about 400 English periodicals, all from India, ranging from popular magazines to scholarly journals. It also serves as current bibliography for books published in India and elsewhere, since it gives full data whenever a book review is cited.

The Bibliography of Asian studies, published since 1956 by the Association for Asian Studies, is a yearly index to books, journal articles, and essays in selected composite works. About one-third of the material is on South Asia. It generally appears three or four years after the year indexed.

There are three current bibliographical sources for Indian books, all providing access by author, title and subject: BEPI: Bibliography of English publications in India (Delhi; D.F.K. Trust, 1977-, latest edition 1979); Indian books in print, 1979 (Indian Bibliographies Bureau, 1979; previous editions for 1972, 1973); Indian books; an annual bibliography, compiled by Ram Gopal Prasher, and others (Today and Tomorrow's, latest edition, for 1974-75, published 1975).

The ICSSR journal of abstracts and reviews, published by the Indian Council of Social Science Research, specializes in sociology, anthropology and related fields. It offers extensive reprints of book reviews from major journals, in addition to abstracts of journal articles. The Indian Book Review Digest, edited by Brahma Chaudhuri (University of Alberta, 1979-), undertakes to index all relevant reviews for the year covered, with short excerpts and/or summaries. Of the numerous review newsletters published by booksellers, the most objective is South Asia in review, from South Asia Books in Columbia, Missouri.

## Guides to Theses and Dissertations

Doctoral dissertations on South Asia, 1966-1970; an annotated bibliography, edited by Frank Joseph Shulman (U. of Michigan, Center for South and Southeast Asian Studies, 1971), lists 1301 works accepted by universities in North America, Europe and Australia. Shulman also edits Doctoral dissertations on Asia, a semi-annual index published since 1975 by the Association for Asian Studies. Theses on Indian subcontinent (1877-1971), compiled by Krishan Gopal lists dissertations on South Asian topics from universities in the U.K., U.S., Canada and Australia. Its utility is diminished by abundant errors. The Association of Indian Universities has published a Bibliography of doctoral dissertations; social science and humanities. Four volumes published to date cover the period 1957-1977. Doctoral dissertations on Pakistan, compiled by Muhammad Anwar (Islamabad, National Commission on Historical and Cultural Research, 1976), indexes works accepted by foreign universities up to 1975.

## Guides to Governmental Publications and Statistics

Two general guides to government publications are useful in locating South Asian materials: the New York Public Library's forty-volume Catalog of government publications in the Research Libraries (G.K. Hall, 1972), and Bibliographic guide to government publications (G.K. Hall, 1974-, latest edition 1979). Rajeshwari Datta has compiled a Union catalogue of the government of Pakistan publications held by libraries in London, Oxford and Cambridge (Mansell, 1967). Similarly titled catalogues have been prepared for India by Datta (Mansell, 1971) and for Ceylon by Teresa Macdonald (Mansell, 1970).

The recent editions of the Census of India consist of hundreds of volumes of statistics, analysis and monographs on various topics. The 1981 Census has just been completed. 1971 Indian Census publications (Office of the Registrar General, 1975) shows the range and types of publications available. The Bibliography of Census publications in India (Office of the Registrar General, 1972) covers all works produced since the founding of the Census in 1872. Although none rivals the Census of India in size and scope, all of the other nations of South Asia conduct regular censuses. A guide to these is the International population census bibliography; Asia (Population Research Center, U. of Texas, 1966).

The Gazetteers are a distinctively South Asian genre. National, provincial and district gazeteers were first compiled in British India and have been updated and published in new editions by the governments of India, Pakistan and Bangladesh. They are compendia of geographic, historical, ethnographic and economic information.

Virendra Kumar's Committees and Commissions in India, 1947 73 (D.K. Publishing House, 1975-) is a guide to the reports of these bodies, dealing with a wide variety of social, economic and cultural problems. Information includes members, terms of reference (i.e. the goals of the committee), contents, and an extensive or complete citation of the recommendations. Nine volumes, covering 1947-69 have appeared as of early 1981. Similar but briefer reports for British India can be found in "Reports of committees and commissions" volume 2, part 6 of Annotated bibliography on the economic history of India, 1500 AD to 1947 AD (Gokhale Institute of Politics and Economics, 1977).

Statistical abstracts, handbooks, pocketbooks, etc. are issued in profusion by governments at all levels. Some general guides are A.T.M. Zahurul Huq, Sources of statistics on Bangladesh: a guide to researchers (Bureau of Economic Research, U. of Dacca, 1978), and Sources of economic and social statistics of India, by Moonis Raza, Shafeeq Naqvi and Jagannath Dhar (New Delhi, Eureka Publications, 1978) India; a reference annual (Ministry of Information and Broadcasting, 1953-) and the Times of India directory and yearbook (Times of India, 1914-) are useful sources of current statistics and other social sciences data.

## Guides to South Asian Economy and Society

For keeping up with scholarly research as well as current events, there is no better journal than the Economic and political weekly from Bombay. In addition to news, editorial comment and book reviews, the Weekly carries academic articles by top scholars, especially in its twice-yearly special issues. Asian survey from Berkeley is likely to have the most timely scholarly analysis of important South Asian events. Internationales Asienforum from Munich is strong on contemporary political, social and economic problems.

A large part of South Asian economic research is conducted by several large research institutes. In India the largest is the National Council of Applied Economic Research; its publications include regional transport surveys, traffic surveys of Indian ports and a series of Techno-economic surveys (1959-67) for most of the Indian states. The privately established Gokhale Institute of Politics and Economics in Poona produces the leading economic journal, Artha vijnana.

The Indian economic review and Indian journal of economics are published by Delhi and Allahabad Universities. The Pakistan and Bangladesh Institutes of Development Economics publish the Pakistan development review and Bangladesh development review. Marga Institute in Colombo publishes Marga; the other leading Sri Lanka economic journal is Central Bank of Ceylon:staff studies. In India, the Planning Commission publishes the Five-Year Plans, together with evaluative reports; parallel agencies in the other nations issue reports on national development plans.

Vimal Rath's Index of Indian economic journals, 1916-1965 (Orient Longman, 1971) covers thirty-two journals. The quarterly Index to Indian economic journals, published between 1968 and 1972, indexed 108 journals. The six volumes of the ICSSR-sponsored A survey of research in economics (Allied, 1975-78) cover methodology, macroeconomics, agriculture, industry and econometrics. Akhtar H. Siddiqui's Economic planning in Pakistan: a select bibliography (Pak Publishers, 1970) and Alauddin Talukder's Bangladesh economy: a select bibliography (Bangladesh Institute of Development Economics, 1976) are the latest examples of the numerous economic sources produced by these two bibliographers.

All of the South Asian nations are primarily agricultural. The Indian Council of Agricultural Research publishes the Indian journal of agricultural economics, for which R.N. Sharma has compiled a cumulative index, Thirty years of Indian journal of agricultural research ... v. i-xxx, 1946-1975 (Concept, 1977). Agriculture: a bibliography, edited by Satyaprakash (Indian Documentation Service, 1977) indexes 172 Indian journals. There are two excellent guides to Indian agricultural statistics: Agricultural statistics in India by P.C. Bansil (Arnold Heinemann India, 2nd rev. edn. 1974), and Statistical sources on Indian agriculture, by D.R. Kalia, M.K. Jain, and T.D. Guliani (Marwah, 1978). An immense body of data on India was collected in the World Agricultural Census 1970-71. The major reports have been published by individual state governments; they are summarized in All-India report on agricultural census, 1970-71, by I.J. Naidu (Dept. of Agriculture, 1975). For Pakistan and Bangladesh, numerous bibliographies have been produced by the Academies of Rural Development in Peshawar and Comilla, and by other agencies. The most recent and valuable are Edward J. Clay's A select bibliography on agricultural economics and rural development, with special reference to Bangladesh (Bangladesh Agricultural Research Council, 1977, with supplement, 1978), and K.M. Bhatti's Bibliography on rural development in Pakistan (Pakistan Academy for Rural Development, 1973). For Sri Lanka there is A bibliography of socio-economic studies in the agrarian sector of Sri Lanka, compiled by W. Ranasinghe, S.M.K. Mileham, and C. Gunatunga (Colombo: Agrarian Research and Training Inst., 1977).

The classic study of South Asia development is Gunnar Myrdal's Asian drama: an enquiry into the poverty of nations (20th Century Fund, 1968); in the intervening 13 years, little has happened to alter the validity of Myrdal's pessimistic analysis. A very recent analysis of India's development experience in its social and political context is Francine Frankel's India's political economy, 1947-1977;the gradual revolution (Princeton U. Press, 1978). In his Sarvodaya:the other development (Vikas, 1980) Detlef Kantowsky evaluates the experience of India and Sri Lanka with an alternative approach based on Gandhian social and economic ideas.

The most essential reference work on South Asian society is Anthropological bibliography of South Asia, edited by Elizabeth Fürer-Haimendorf and Helen Kanitkar (Mouton, 1958, 1964, 1970, 1976); the four volumes published to date cover scholarship up to 1969. The entries, both published works and research in progress, are arranged according to twenty sociocultural regions. The three volumes of the ICSSR-sponsored A survey of research in sociology and social anthropology (Popular Prakashan, 1972-74) provide trend reports and bibliographies on twenty-six subject areas. Danesh Chekki's The social system and culture of modern India: a research bibliography (Garland, 1975) lists 5487 items in a classified scheme taken from Sociological Abstracts with slight modifications. This scheme does not give enough recognition to the special features of Indian society, such as caste, and the book also lacks indexes.

That peculiar institution of South Asian society, the caste system, has attracted the interest of Westerners from the first contacts. In his recent controversial classic, Homo hierarchicus; the caste system and its implications (U.of Chicago Press, rev. edn. 1980) Louis Dumont regards hierarchy as the quintessence of Indic society, in contrast to Western egalitarianism, and he relates caste to kingship, religion, and other features of the civilization. In the preface to the revised edition, the author answers a decade of criticism. Caste in contemporary India; beyond organic solidarity by Pauline Kolenda (Benjamin/Cummins, 1978) includes a lengthy and thorough review of the literature.

Compared to Europe, North America, or East Asia, South Asian society presents an image of extreme fragmentation; not only are there wide geographic differences in society and culture, but even in the same local area one finds diverse groups, such as "Caste Hindus", Scheduled Castes or "Untouchables", tribal peoples, Muslims, Eurasians and other linguistic, ethnic, and religious minorities. Clarence Maloney's Peoples of South Asia (Holt, Rinehart and Winston, 1974) is a general introduction emphasizing the diversity of social and cultural groups. Essays on the Scheduled Castes, with a forty five page bibliography, are found in The Untouchables in contemporary India, edited by J. Michael Mahar (U. of Arizona

Press, 1972). Stephen Fuchs' The aboriginal tribes of India (St. Martin's Press, 1973) is a subcontinent-wide survey; the extensive bibliographies of each regional chapter are brought together by the author index. Islam in southern Asia; a survey of current research, edited by Dietmar Rothermund (Franz Steiner, 1975) deals with the social and political status of Muslims. Family, kinship and marriage among Muslims in India (Manohar, 1976) and Caste and social stratification among Muslims in India (Manohar, 2nd edn.,1978), both edited by Imtiaz Ahmad, are part of a projected four-volume series on Muslim life. A social category neglected by scholarship until quite recently is covered by Carol Sakala's Women of South Asia; a guide to resources (Kraus International, 1980); the 4629-item classified and annotated bibliography is supplemented by essays on archives, libraries and other resources. Jagdish Saran Sharma's India's minorities; a bibliographical study (Vikas, 1975) covers Anglo-Indians and seven religious groups. Linguistic groups are discussed and mapped by Roland J.-L. Breton in his Atlas géographique des langues et des ethnies de l'Inde et du subcontinent... (Presses de l'université Laval, 1976)

Village studies, a common and important type of research, vary widely in approach and content; while most attempt some overall description of the community, one may emphasize religion and another irrigation projects. This diversity is admirably dealt with in Village studies: analysis and bibliography, v.1, India 1950-1975 (Inst. of Development Studies, U. of Sussex, 1976). Each entry includes sixty coded fields specifying subjects covered and the type of information available.

## General Atlases

A historical atlas of South Asia, edited by Joseph E. Schwartzberg and others (U. of Chicago Press, 1978) is a far more comprehensive reference work than its title indicates. Its 682 maps vividly present information on every field of the social sciences and humanities, from prehistory to the 1970's. In addition, it has over 120 pages of text and a bibliography of over 3500 items.

The National atlas of India (National Atlas Organisation, 1977) includes 126 maps with accompanying text, showing physical features, administrative divisions and demographic and economic data. The Office of the Registrar General has published a number of atlases as part of the Census of India: in the 1961 and 1971 censuses, national and state atlases constitute part 9 of each volume. The Census centenary atlas (1974) is a convenient summary of demographic and social data prepared for the World Population Year.

Shorter atlases convenient for reference or classroom use are South Asia in maps by Robert C. Kingsbury (Denoyer-Geppert, 1969), and India in maps by Ashok K. Dutt, S.P. Chatter-

-jee, and M. Margaret Geib (Kendall/Hunt, 1976). For Sri Lanka, there is A concise atlas geography of Ceylon (Colombo: Atlas and Map Industries, 1971).

## Geographical Bibliography

The standard Western geographical indexes contain material on South Asia. These include Geo Abstracts, Bibliographie géographique internationale, and the Current geographical publications. However, the proportion of studies on South Asia is very small, and my own survey indicates that many books and articles from the subcontinent do not appear, either because they were not reported or because they were deemed to be deficient in scholarship.

B.L. Sukhwal's South Asia: a systematic geographical bibliography (Scarecrow Press, 1974) indexes over 10,000 books, articles and dissertations by country and topic. Sukhwal has also compiled A systematic geographical bibliography on Bangla Desh (Monticello, IL: Council of Planning Librarians, 1973). Mushtaqur Rahman has compiled a Bibliography of Pakistan geography, 1947-73 (Council of Planning Librarians, 1974). Sections on the geography of Sri Lanka and Nepal can be found in the Goonetileke and Boulnois works cited above in the sections on general bibliographies.

S.P. Chatterjee's Progress of geography, and its supplement, Progress of geography, 1964-1968 (Indian Science Congress Association; 1964, 1968) offer topic-by-topic histories of geographical research in India, with bibliographies. The ICSSR-sponsored A survey of research in geography (Popular Prakashan, 1972) consists of trend reports by experts in many subject areas, with bibliographies. Its supplement is A Survey of research in geography, 1969-72 (Allied, 1979).

## Geographical Journals

The most important journals are those of the national level geographical societies. In India, these are the National geographical journal of India (Varanasi, 1955-), the Geographical review of India (Calcutta, 1936-) and the Indian geographical journal (Madras, 1926-). Pakistan has the Pakistan geographical review (Lahore, 1942-) and Bangladesh the Oriental geographer (Dacca, 1957-). The Nepalese Geographical Society publishes the annual Himalayan review (Kirtipur, 1968-).

There are many minor geographical journals, put out by regional societies or university departments. Examples are the Deccan geographer from Hyderabad, and Uttar Bharat bhoogol patrika (North India geographical journal) published at Gorakhpur University. These can be found in the International list of geographical serials, edited by Chauncy D. Harris and Jerome D. Feldman (U. of Chicago Dept. of Geography, 3rd edn., 1980). Including vernacular and discontinued serials, this source lists seventy-one titles from South Asia, of which fifty-three are Indian.

## General and Regional Works in Geography

O.H.K. Spate and A.T.A. Learmouth, India and Pakistan; a general and regional geography, with a chapter on Ceylon by B.L. Farmer (Methuen, 3rd edn., 1967) is by all odds the most important and comprehensive study to date on the geography of the subcontinent. The other major recent study covering most of the subcontinent is India; a regional geography (National Geographic Society of India, 1971), by R.L. Singh and others. This work is distinguished by excellent, detailed coverage of twenty-eight physiographic regions and their subregions. A stimulating new collection of essays, edited by David E. Sopher, An exploration of India: geographical perspectives on society and culture (Cornell U. Press, 1980) is marked by innovative use of statistics and suggestive of new concepts and techniques. For example, in his lead article Bharat L. Bhatt introduces the notion of "folk regions" to the continuing discussion of regionalization.

B.L.C. Johnson has written several introductory texts on South Asian geography, all published by Heinemann Educational Books: South Asia; selective studies in the essential geography of India, Pakistan, and Ceylon (1969); India; resources and development (1979); Bangladesh (1975); and Pakistan (1979). Two excellent short introductions to Pakistan are Kazi S. Ahmad, A geography of Pakistan (Karachi: Oxford U. Press, 3rd edn., 1972), and K.U. Kureshy, A geography of Pakistan (Karachi: Oxford U. Press, 1977). For Sri Lanka, the only recent general geographical study is Manfred Dömros, Sri Lanka: die Tropeninsel Ceylon (Wissenschaftliche Buchgesellschaft, 1976). Pradyumna Prasad Karan has written two introductions to the Himalayan kingdoms: Geography of Nepal: physical, economic, cultural and regional (U. of Kentucky Press, 1970) and Bhutan: a physical and cultural geography (U. of Kentucky Press, 1967).

Many studies are written based on political units; this procedure is not without justification, since the states of India and provinces of Pakistan are in general based on cultural-linguistic regions. By far the largest quantity of regional studies is published in the journals published in the subcontinent, as well as in special volumes of the censuses, such as Economic and socio-cultural dimensions of regionalisation..., from the 1971 Census of India (Office of the Registrar General, 1972). The books cited here are merely examples and by no means comprise an exhaustive list of monographs.

The series India - the land and people, published by the National Book Trust in New Delhi, includes a number of high-quality geographic monographs on the states.

Geography of Mysore, by Rameshwar Prasad Misra (NBT, 1973)

Geography of Maharashtra, by Chandrasekhar Dundiraj Deshpande (NBT, 1971)

Geography of Gujarat, by Kamal Ramprit Dikshit (NBT, 1970)

Geography of Rajasthan, by Vinod Chandra Misra (NBT, 1969)

Geography of Orissa, by Bichitrananda Sinha (NBT, 1971)

Geography of Jammu & Kashmir, by A.N. Raina (NBT, 1971)

Geography of Uttar Pradesh, by Angelo Rajkumar Tiwari, (NBT, 1971)

Geography of West Bengal, by Subhas Chandra Bose (NBT,1968)

Geography of Assam, by Hari Prasanna Das (NBT, 1970)

B. Arunachalam has written Maharashtra: a study in regional setting and resource development (Bombay: A.R. Sheth, 1967). Two important studies of the lower Indus valley are Mushtaqur Rahman, A geography of Sind Province, Pakistan (Karachi Geographic Association, 1975), and Maneckji Bejanji Pithawala, Physical and economic geography of Sind... (Sindhi Adabi Board, 1959). Other regions of the Northwest are covered in Hiroshi Ishida, Dynamic regional studies of the Punjab, India (U. of Hiroshima, 1975), David Dichter's The Northwest Frontier of West Pakistan, a study in regional geography (Oxford U. Press, 1967) and The Valley of Kashmir, a geographical interpretation, by Moonis Raza and others (Vikas, 1978). Enayat Ahmad's Bihar: a physical, eonomic and regional geography (Ranchi U., 1965) is an excellent though somewhat outdated study of one of India's poorest states. Geography of the Himalayas, by Subhas Chandra Bose (National Book Trust, 1972) covers the Indian portions of the mountain rim. Parmanand Lal's Andaman Islands: a regional geography (Anthropological Survey of India, 1976) describes the Indian union territory in the Bay of Bengal.

Topical Works in Geography

Several monographs in the India - the land and people series deal with physical geography: Physical geography of India, by Charles S. Pichamuthu (NBT, 1967), and Rivers of India by S.D. Misra (NBT, 1970). The South Asian climate is described in The monsoons, by Prosad Kumar Das (NBT, 1970).

Two excellent recent economic geographies of India are Subhas Chandra Bose, Economic geography of India (Annapurna, 1974), and Manoranjan Chaudhuri, An economic geography of India (Oxford & IBH, 1971). Nafis Ahmad has written A new economic geography of Bangladesh (Vikas, 1975); his earlier, larger An economic geography of East Pakistan (Oxford U. Press, 1968) is still worth consulting for basic information. A recent economic geography of Nepal is Nepal: Raum, Mensch

Wirtschaft by Wolf Donner (Otto Harrassowitz, 1972).

Recent works on agricultural geography include Geography of India: agricultural geography, by Caturbhuj Mamoria (Shiva Lal Agarwala, 1975) and Jasbir Singh's An agricultural geography of Haryana (Vishal, 1976). Jasbir Singh also edited An agricultural atlas of India: a geographical analysis (Vishal, 1974). The role of regions in economic development is analysed by R.P. Misra in his Regional planning; concepts, techniques, policies and case studies (U. of Mysore, 1969) and by Enayat Ahmad and Devendra Kumar Singh in their Regional planning, with particular reference to India (Oriental, 1980). Bichitranath Sinha's Industrial geography of India (World Press, 1972) discusses optimal and actual locations of industries.

Ashish Bose has edited two basic bibliographies on urbanisation, both covering research after 1947: Urbanisation in India; an inventory of source materials (Academic Books, 1970), and Bibliography on urbanization in India, 1947-1976 (Tata McGraw-Hill, 1976). John Van Willigen's two volume The Indian city: a bibliographic guide to the literature on urban India (Human Relations Area files, 1979) lists 3801 items by city or area. Million cities of India, edited by R.P. Misra (Vikas, 1978), offers essays on the nation's nine largest urban centres.

Rural settlements are a special area of interest of the Varanasi school, centred around Banaras Hindu University and the National Geographical Society of India under the leadership of R.L. Singh. Recent examples of their work in this field are two volumes edited by Singh: Geographic dimensions of rural settlements (NGSI, 1976) and Readings in rural settlement geography (NGSI, 1975).

B.L. Sukhwal's India: a political geography (Allied, 1971) analyses the role of geographic factors in the nation's party politics and international relations, while V.A. Janaki's Some aspects of the political geography of India (Maharaja Sayajirao U. of Baroda, 1977) offers essays on both historical and contemporary themes. Pakistan; a political geography, by Ali Tayyeb (Oxford U. Press, 1966), discusses the problems of national integration that led to the subsequent division of the nation.

In addition to Joseph Schwartzberg's Historical atlas cited above, the much briefer Historical atlas of the Indian peninsula, by Cuthbert Colin Davies (Oxford U. Press, 2nd edn. 1959), remains a standard reference work. An atlas of the Mughal Empire; political and economic maps... (Oxford U. Press, 1981) is a new work edited by Irfan Habib, the leading economic historian for the period. Two studies published by the Duke University Program in Comparative Studies on Southern Asia analyse the historical and political role of regions: Regions and regionalism in South Asian studies, edited by Robert I. Crane, (1969), and Realm and region in traditional India, edited by Richard G. Fox (1979). For the earlier periods, Bimala Churn Law's Historical geography of ancient India is a dictionary of place names found in classic texts.

An increasingly important area of cultural geography is the geography of religion, especially the study of the circuit of pilgrimage centres that has constituted one of the great unifying factors in Indic civilization throughout history. Considerable material on South Asia is found in two general works: David E. Sopher's Geography of Religions (Prentice-Hall, 1967), and Historical atlas of the religions of the world, edited by Sopher and Isma'īl Rāgī al Fārūqī (Macmillan, 1974). The pattern of Hindu sacred centres is analysed by Surinder Mohan Bhardwaj in his Hindu places of pilgrimage in India; a study in cultural geography (U. of California Press, 1973). An example of one of the many studies of specific centres is Cultural tradition in Puri; structure and organization of a pilgrim center, by N. Patnaik (Indian Institute of Advanced Study, 1977).

Chapter Eight

# U.S.A.

J.H. Wheeler

Geographical information on the United States is voluminous, but the sources are very diffuse. Pertinent writings by professional geographers and by workers in cognate fields are scattered through a formidable mass of articles, treatises, textbooks, encyclopaedias, and reports. Much valuable material is contained in dissertations and theses presented by candidates for graduate degrees. The volume of statistical, cartographic, and photographic material is immense. And not to be overlooked are the writings of journalists and novelists, some of which richly illuminate the American scene.

How is the searcher to find a way through this welter of fact and commentary? The task is not eased by the outdated character of much of the material. The United States has seen enormous changes in recent decades, and geographical writing often has not kept pace. Information in older works must be used with discretion, and recent works afford a very uneven coverage by place and topic. Major outlines and trends can be discerned from the available material, but the searcher often will find a provoking lack of specialized in-depth analyses.

Attention will be focused initially on bibliographies, serials, and other sources that contain important material on the United States along with material on other parts of the world. A discussion of sources centring explicitly on the geography of the United States as a whole will follow, and will include summations of subfields such as historical geography or physical geography. Finally, there will be guidance to sources of a regional character, organized under the headings of Northeast, Middle West, South, and West.

## International Sources that include the United States

Elaborate lists of citations to publications on the United States in geography and closely-related fields will be found in the *Research Catalogue of the American Geographical Society*, 15 vols. (Hall, 1962), its *First Supplement Regional 1962-1971*, Vol.1 (1972), its *Second Supplement 1972-1976*, Vol.2, *Regional* (1978), and *Current Geographical Publications*

(AGS and University of Wisconsin-Milwaukee Library, 10 issues a year). Numerous citations can also be found in the Bibliographie Géographique Internationale (Centre National de la Recherche Scientifique, Paris, annual, 1891-1976 and quarterly since 1977); Geo Abstracts (University of East Anglia; annotated; 6 issues a year in 7 series A-G); and New Geographical Literature and Maps (Royal Geographical Society; semiannual; ceased publication in 1981). A sizable share of the references on the United States cited in the foregoing sources are articles or monographs appearing in geographical serials. Particularly important among the latter are serials published in the United States, as listed on pp. 129-37 of C.D. Harris, Annotated World List of Selected Current Geographical Serials (University of Chicago, Department of Geography Research Paper 194, 1980). Key geographical articles are especially numerous in three quarterly journals: Annals of the Association of American Geographers, Geographical Review, and Economic Geography. Indexes have been published by the respective journals and by Snipe International, Manitou Springs, Colorado. Other serials of broad utility include the Journal of Geography, Professional Geographer, Focus, Landscape, Ecumene, Geographical Perspectives, Geographical Survey, Geographical Bulletin, Antipode, East Lakes Geographer, AAG Resource Papers for College Geography, University of Chicago Department of Geography Research Papers, and University of California Publications in Geography. Serials with a strong topical or regional focus are cited under appropriate headings later in the chapter. Beyond the serials cited above, a wide scatter of occasional professional articles on the geography of the United States is published in numerous geographical journals in English, French, German, or other languages (for example, Geographical Magazine, Geography, IBG Transactions, Geoforum, GeoJournal, Annales de Géographie, Geografiska Annaler). A world list of journals and other serials is afforded by C.D. Harris and J.D. Fellman, International List of Geographical Serials (3rd edn. University of Chicago, Geography Research Paper 193, 1980). Not to be disregarded, especially in regard to maps and photographs, is that phenomenally successful mass-circulation monthly, the National Geographic. A kindred magazine, Geo, also publishes some articles on the United States. Articles containing information of geographical import appear in a practically endless number of scholarly journals in cognate fields or in mass-circulation magazines and newspapers such as the Scientific American, Natural History, Fortune, the New Yorker, the New Republic, Harper's, the weekly news magazines, and the New York Times, Christian Science Monitor, and Wall Street Journal. The searcher for information in cognate fields and general-readership publications will be aided by such general indexes as the Reader's Guide to Periodical Literature or the Social Sciences Index.

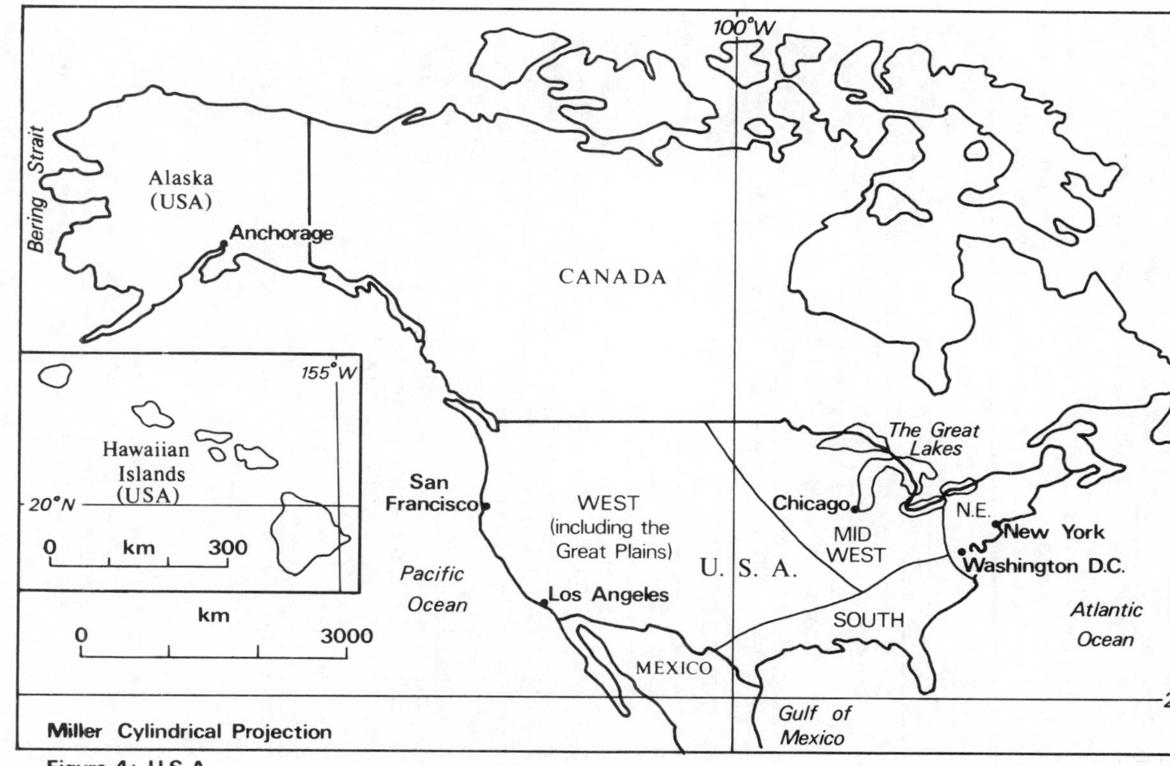
100°W
Bering Strait
Alaska (USA)
Anchorage
CANADA
155°W
Hawaiian Islands (USA)
20°N
0 km 300
km
0 3000
San Francisco
Pacific Ocean
WEST (including the Great Plains)
Los Angeles
U. S. A.
MEXICO
Chicago
MID WEST
The Great Lakes
N.E.
New York
Washington D.C.
SOUTH
Atlantic Ocean
Gulf of Mexico
25°N
Miller Cylindrical Projection

Figure 4: U.S.A.

Many serials published by scholarly societies include lists, notices, and/or reviews of new books; for the United States a particularly comprehensive source is the AAG Newsletter (bi-monthly). The searcher may also consult Current Geographical Publications, Section Four, which lists accessions to the AGS Collection of the University of Wisconsin-Milwaukee Library.

Important geographical information on the United States can often be secured from doctoral dissertations and master's theses. Periodic lists of those accepted by United States graduate departments of geography are published in the Professional Geographer, as well as by individual departments in AAG, Guide to Graduate Departments of Geography in the United States and Canada (annual). Comprehensive lists include C.E. Browning, A Bibliography of Dissertations in Geography: 1901 to 1969- American and Canadian Universities (University of North Carolina at Chapel Hill, Department of Geography, Studies in Geography 1, 1970) and M.M. Stuart, A Bibliography of Master's Theses in Geography: American and Canadian Universities (Tualatin, Oregon: Geographic and Area Study Publications, 1973). Dissertations and theses are available on microfilm from University Microfilms, Ann Arbor, Michigan; some are published in research series of their respective universities, but most are unpublished.

Innumerable world atlases contain valuable regional and topical maps of the United States; a few recent examples are Goode's World Atlas (Rand McNally, 16th edn., 1981), The New International Atlas (Rand McNally, 1980), and The Times Atlas of the World (Times Books, 1981). Atlases centred on the United States are cited in the following section. For world bibliographies of atlases and maps, see the Bibliographie Cartographique Internationale (Comité National Francais de Géographie; International Geographical Union, Paris; ceased publication 1975); the AGS Index to Maps in Books and Periodicals, 10 vols. (Hall, 1968), together with the First Supplement (1971) and Second Supplement (1976); Research Libraries of the New York Public Library and the Library of Congress, Bibliographic Guide to Maps and Atlases: 1979 (Hall, first annual volume, 1980); Current Geographical Publications; New Geographical Literature and Maps; the Bulletin of the Geography and Map Division, Special Libraries Association; and GeoKartenbrief (GeoCenter Internationales Landkartenhaus, Stuttgart; monthly).

Statistical and technical publications of the United Nations, Population Reference Bureau, and kindred agencies include data for the United States, as do yearbooks such as the Britannica Book of the Year or the Statesman's Year-Book. Statistical sources specifically for the United States will be cited in the following section. Finally, one must not forget the large amount of geographical information contained in the standard encyclopedias. Many relevant articles were written by professional geographers, for example, P.F. Lewis, "The Natural Landscape", and W. Zelinsky, "Patterns of Settlement and Cultural Development," in the article "United States of America", Vol.18, Encyclopaedia Britannica, 15th edn., (1974). The one-volume New Columbia Encyclopedia (Columbia Univ. Press, 1975) is to be especially recommended for its remarkable complement of essential information available in a minimal space. It may be noted that this source provides a selection and updating of place data from the older Columbia Lippincott Gazetteer of the World (1962).

## The United States as a Whole

We move now to a summation of selected books, articles, and bibliographies on the United States as a whole, together with atlases and maps, U.S. Government documents (including statistical sources), and sources in cognate fields. For general geographical information on the entire United States, one may profitably turn to recent editions of university-level textbooks, including J.H. Peterson, North America: A Geography of Canada and the United States (Oxford University Press, 6th edn., 1979); R.S. Thoman, The United States and Canada: Present and Future (Merrill, 1978); and C.L. White, E.J. Foscue, and T.L. McKnight, Regional Geography of Anglo-America (Prentice-Hall, 5th edn., 1979). All have bibliographical citations (books and articles) comprising hundreds of titles. Also useful, though lacking bibliography, is S.S. Birdsall and J.W. Florin, Regional Landscapes of the United States and Canada (Wiley, 2nd edn., 1981). Other textbooks in English include R. Estall, A Modern Geography of the United States (Penguin, 1976); O.P. Starkey, J.L. Robinson, and C.S. Miller, The Anglo-American Realm (McGraw-Hill, 2nd edn., 1975); G.H. Dury and R.S. Mathieson, The United States and Canada (Heinemann, 2nd edn., 1976); J.W. Watson, North America: Its Countries and Regions (Longmans,1968); W.R. Mead and E.H. Brown, The United States and Canada: A Regional Geography (Hutchinson Educational, 1962); and an older classic, J.R. Smith and M.O. Phillips, North America (Harcourt, 1942). J.F. Hart, ed., Regions of the United States (Harper, 1972) is a valuable collection of articles published originally as the June, 1972, issue of the AAG Annals. Specific articles from this source are cited under the appropriate regions in the latter part of this chapter. See also R.E. Dickinson, Environments of America (Vantage Press, 1974). A selected list of references on the United States totalling some 250 items accompanies two chapters on Anglo-America in J.H. Wheeler, Jr., J.T. Kostbade, and R.S. Thoman, Regional Geography of the World (Holt, 3rd edn., 1969), and an abridgement of the list plus some 150 new references accompanies the 3rd updated edition of this text (1975). A recent synthesis in French is J.H. Wheeler, Jr., "Les Etats-Unis," pp. 1085-1410 in M.J.B. Delamarre et al., Géographie Régionale 2 (Gallimard, 1979). A bibliography of about 170 items, mainly in English but including some citations in French or German, is appended. See also J.W. Watson, "Geography and the Development of the U.S.A.", in D. Welland, ed., The United States: A Companion to American Studies (Methuen, 1974).

H.F Raup, "The United States in Professional Geographic Literature," Annals AAG 46 (1956), and A.H. Doerr, "The United States in Professional Geographic Literature: A Recent View," Professional Geographer 16 (1964) are summations that utilize dot maps to indicate studies of places published in selected journals and doctoral dissertations. S.B. Cohen, ed., Problems and Trends in American Geography (Basic Books,1967) and R.S. Platt, Field Study in American Geography (Univ. of Chicago, Geography Research Paper 61, 1959) provide a miscellany of substantive studies along with commentaries on theory and method. A useful recent source in German is H. Blume, Wissenschaftliche Landerkunden Band 9/1: USA, I. Der Grossraum in Strukturellem Wandel; Band 9/II: USA, II. Die Regionen der USA (Wissenschaftliche Buchgesellschaft, Darmstadt). For citations to older works, see W.L.G. Joerg, "The Geography of North America: A History of Its Regional Exposition," Geographical Review 26 (1936).

Statistical publications of first importance for general reference include U.S. Bureau of the Census, Statistical Abstract of the United States (GPO, annual) and, from the same source, the Pocket Data Book, USA (GPO, biennial; 6th edn., 1980, data for 1979), the County and City Data Book (GPO, from 1949; every 5th year from 1962), and the State and Metropolitan Area Data Book, 1979 (GPO, 1980). For more detailed statistical reference, consult the innumerable sources listed in the Monthly Catalog of United States Government Publications (GPO). Some statistical sources of a topical nature are cited later in this chapter.

A broad spectrum of maps is contained in U.S. Geological Survey, The National Atlas of the United States of America (GPO, 1970). See also the Commercial Atlas and Marketing Guide (Rand McNally, annual); D.K. Adams, S.F. Mills, and H.B. Rodgers, An Atlas of North American Affairs 2nd edn., (Methuen, 1979); and the many maps published by the National Geographic Society as supplements to the National Geographic Magazine. Topographic and other maps issued by the Geological Survey are cited in New Publications of the Geological Survey (U.S. Department of the Interior, monthly), and a general guide to the Survey's cartographic products is available in U.S. Geological Survey, Maps for America (GPO, 1979). Cartographic sources of a topical or regional nature are cited under appropriate headings below.

Annotated lists of books on the United States in many fields appear in U.S. Library of Congress, Reference Department, A Guide to the Study of the United States of America: Representative Books Illustrating the Development of American

Life and Thought (GPO, 1960) and its Supplement, 1956-1965 (GPO, 1976). The TIME-LIFE Library of America, 13 vols. (Time, Inc., 1967-1969) is a massive source of text, photos, maps, and bibliography, written for general readers. The volumes are organized by groups of states. Another source containing much perceptive writing for the general reader is H. Moseley, ed., The Romance of North America (Houghton Mifflin, 1958).

General works on the United States by journalists include J. Gunther, Inside U.S.A. (Harper, 1947) and N.R. Peirce, The Megastates of America (Norton, 1972). Works of fiction are cited in O.W. Coan and R.G. Lillard, America in Fiction: An Annotated List of Novels That Interpret Aspects of Life in the United States, Canada, and Mexico (Pacific Books, 5th edn., 1967). See also J. Conron, ed., The American Landscape: A Critical Anthology of Prose and Poetry (OUP, 1974).

References in Historical and Topical Geography

Historical geography and the major topical fields of geography are surveyed by the contributors to P.E. James and C.F. Jones, eds., American Geography: Inventory and Prospect (Syracuse Univ. Press for the Association of American Geographers, 1954). Hundreds of titles of books and articles on the United States are cited in the chapter bibliographies. Individual chapters cover the field of geography, the regional concept and method, historical geography, population, settlement, urban geography, political geography, resources, economic geography, agriculture, minerals, manufacturing, transportation, climatology, geomorphology, soils, water, the oceans, plant geography, animal geography, medical geography, physiological climatology, military geography, field techniques, aerial and satellite photograph interpretation, and cartography. This volume is invaluable for commentaries and voluminous citations up to the early 1950's. A more recent and highly useful annotated compilation of book titles is A Geographical Bibliography for American College Libraries (AAG Commission on College Geography, Publication 9, 1970). Organized topically and regionally, it covers the entire field of geography, but includes a large number of references on the United States. The subsections of this chapter that immediately follow will discuss, in order, the topics of historical geography; physical geography and environmental studies; political, cultural, and social geography; economic geography and urban geography.

Historical Geography

The references accompanying A.H. Clark's chapter on historical geography in American Geography: Inventory and Prospect (cited above) are an important bibliographical source. See also D.R. McManis, Historical Geography of the United States: A Bibliography (Division of Field Services, Eastern Michigan Univ., 1965). More recent summations of work in historical geography, accompanied by voluminous reference footnotes, include H.C. Prince, "Three Realms of Historical Geography: Real, Imagined, and Abstract Worlds of the Past," in *Progress in Geography* 3 (1971); D.W. Meinig, "The Continuous Shaping of America: A Prospectus for Geographers and Historians," *American Historical Review* 83 (1978); and J.A. Jakle, "Time, Space, and the Geographic Past: A Prospectus for Historical Geography," *American Historical Review* 76 (1971). *The Journal of Historical Geography* (quarterly) contains many articles and reviews on the United States. R.H. Brown, *Historical Geography of the United States* (Harcourt, 1948), is a classic synthesis, though its purview ends with the late nineteenth century. Also highly regarded is Brown's *Mirror for Americans: Likeness of the Eastern Seaboard, 1810* (American Geographical Society, Special Publication 27, 1943; reprinted by Da Capo Press, 1968). An older work still worthy of note is E.C. Semple, *American History and its Geographic Conditions* (Houghton Mifflin, 1903), revised in collaboration with the author by C.F. Jones (Houghton Mifflin, 1933; reprinted by Russell and Russell, 1969). Of great importance are the volumes in the *Andrew H. Clark Series on the Historical Geography of North America* (OUP), several of which are cited under regional headings later in this chapter. See also M. Conzen, advisory editor, "Fashioning the American Landscape," a series of 12 articles by various authors in the *Geographical Magazine* 52 (1979) and 53 (1980).

Other references under the general heading of historical geography include D. Ward, ed., *Geographic Perspectives on America's Past: Readings on the Historical Geography of the United States* (OUP, 1979); J.R. Gibson, ed., *European Settlement and Development in North America: Essays on Geographical Change in Honour and Memory of Andrew Hill Clark* (Univ. of Toronto Press, 1978); H.H. Barrows, *Lectures on the Historical Geography of the United States as Given in 1933*, ed. by W.A. Koelsch (Univ. of Chicago, Geography Research Paper 77, 1962); R.E. Ehrenburg, ed., *Pattern and Process: Research in Historical Geography* (Howard Univ. Press, 1975); W.D. Pattison, *Beginnings of the American Rectangular Land Survey System, 1784-1800* (Univ. of Chicago, Geography Research Paper 50, 1957); N.J.W. Thrower, *Original Survey and Land Subdivision: A Comparative Study of the Form and Effect of Contrasting Cadastral Surveys* (AAG Monograph 4, 1966); and several publications by C.O. Sauer, including "A Geographical Sketch of Early Man in America," *Geographical Review* 34 (1944); *The Early Spanish Main* (Univ. of California Press, 1969); *Sixteenth Century North America* (Univ. of California Press, 1971) and *Seventeenth Century North America* (Netzahaulcoyotl Historical Society, Berkeley, California, 1980).

Some important cartographic and statistical references include L.J. Cappon, "The Historical Map in American Atlases," *Annals AAG* 69 (1979); C.O. Paullin, *Atlas of the Historical Geography of the United States*, ed. J.K. Wright (Carnegie

Institution of Washington and American Geographical Society, 1932); the many historical maps in The National Atlas of the United States of America; H.H. Kagan et al., The American Heritage Pictorial Atlas of United States History (American Heritage, 1966); T.R. Miller, Graphic History of the Americas (Wiley, 1969); J.T. Adams and R.V. Coleman, eds., Atlas of American History (1943; rev. edn. edited by K.T. Jackson, Scribner, 1978); E.W. Fox, Atlas of American History (OUP, 1964);R.D. Sale and E.D. Karn, American Expansion: A Book of Maps (Univ. of Nebraska Press, 1962, 1979); L.J. Cappon et al., eds., Atlas of Early American History: The Revolutionary Era, 1760-1790 (Princeton, 1976); and U.S. Bureau of the Census, Historical Statistics of the United States: Colonial Times to 1970, 2 Parts (GPO, 1975).

A small selection of cognate general works on prehistory and history includes J.D. Jennings, Prehistory of North America (McGraw-Hill, 2nd edn., 1974); B. Bailyn et al., The Great Republic: A History of the American People (Little, Brown, 1977); S.E. Morison, H.S. Commager, and W.E. Leuchtenburg, The Growth of the American Republic, 2 vols. (OUP, 7th edn., 1980); D.J. Boorstin, The Americans, 3 vols.: The Colonial Experience (1958), The National Experience (1965), The Democratic Experience (1973) (Random House; Weidenfeld, and Nicolson); The LIFE History of the United States, 12 vols. (Time, Inc., 1963-1964); A.M. Schlesinger and D.R. Fox, eds., A History of American Life, 13 vols. (Macmillan, 1927-1948); H.S. Commager and R.B. Morris, eds., The New American Nation Series, particularly H. Savage, Jr., Discovering America, Seventeen Hundred to Eighteen Seventy-Five (Harper and Row, 1979); the many volumes in D.J. Boorstin, ed., The Chicago History of American Civilization (Univ. of Chicago Press); and U.S. National Park Service, The National Survey of Historic Sites and Buildings, 11 vols. (GPO, 1967). The number of specialized articles and monographs on American history is truly enormous; for general guidance see F. Freidel, ed., Harvard Guide to American History, rev. edn., 2 vols. (Harvard Univ. Press, 1974) and E. Cassara, History of the United States of America: A Guide to Information Sources (Gale Research Co., 1977). See also T.H. Johnson, ed., The Oxford Companion to American History (OUP, 1966), and issues of the American Historical Review (5 issues a year), Recently Published Articles (American Historical Association, quarterly), Journal of American History (quarterly), and, for the general reader, American Heritage (6 issues a year). Additional historical references are cited under appropriate topical and regional headings in subsequent sections of this chapter.

Physical Geography and Environmental Studies

The most broadly useful volume on the physical geography of the United States is C.B. Hunt, Natural Regions of the United States and Canada (Freeman, 1974). Very fully illustrated, it contains reference lists of books and technical papers. Discussions of physical geography in briefer compass may be gained from the regional geographies cited earlier or piecemeal from general texts on physical geography such as the superbly illustrated volume by A.N. Strahler and A.H. Strahler entitled Modern Physical Geography (Wiley, 1978). Standard works on physiography and geomorphology include N.M. Fenneman's Physiography of Western United States (McGraw-Hill, 1931) and Physiography of Eastern United States (McGraw-Hill, 1938); W.D. Thornbury, Regional Geomorphology of the United States (Wiley, 1965); and A.K. Lobeck, Geomorphology (McGraw-Hill, 1939). Lobeck's book is a finely illustrated general work from which many insights on the United States may be secured. W.W. Atwood, The Physiographic Provinces of North America (Ginn, 1940) is rather dated but still useful. J.A. Shimer, This Sculptured Earth: The Landscape of America (Columbia Univ. Press, 1959) was written for general readers. Indispensable to any student of physiography are the many maps and physiographic diagrams by Fenneman, Lobeck, Raisz, and Hammond, particularly N.M. Fenneman, Physical Divisions of the United States (U.S. Geological Survey, 1946 edn.); A.K. Lobeck, Physiographic Diagram of the United States (Hammond, 1932); E. Raisz, Landforms of the United States (E. Raisz, 130 Charles St., Boston MA; 6th rev. edn., 1957); and E.H. Hammond, Classes of Land-Surface Form in the Forty Eight States, U.S.A. (AAG Annals, Map Supplement 4, 1964; reprinted in The National Atlas of the United States of America and available as a separate sheet from U.S. Geological Survey). See also U.S. Geological Survey, A Set of One Hundred Topographic Maps Illustrating Specified Physiographic Features (1957), and R. Knowles and P.W.E. Stowe, North America in Maps: Topographical Map Studies of Canada and the USA (Longmans,1976). Not to be overlooked is the masterly exhibit of satellite photographs of the United States included in N.M. Short et al., Mission to Earth: Landsat Views the World (Washington, D.C.: National Aeronautics and Space Administration, 1976). Also based on satellite imagery is Photo-Geographic International, Photo-Atlas of the United States (Pasadena CA: Ward Ritchie Press, 1975). Geologic evolution is treated in C.W. Stearn et al., Geological Evolution of North America (Wiley, 3rd edn., 1979); P.B. King, The Evolution of North America (Princeton Univ. Press, rev. edn., 1977); and H.E. Wright and D.G. Frey, eds., The Quaternary of the United States (Princeton Univ. Press, 1965). See also J.S. Shelton, Geology Illustrated (Freeman, 1966), which incorporates fine air photographs of characteristic physical features. For data on minerals, see the publications of the U.S. Bureau of Mines, especially the annual Minerals Yearbook (GPO) and Mineral Facts and Problems (Bulletin 650, 1970; GPO). In recent years the United States has seen an enormous outpouring of information on energy resources; a concise survey for the general reader, including numerous maps and a bibliography, is offered by the National Geographic Special Report, Energy (Washington DC; National Geographic Society, February, 1981). On water resources, see U.S. Department of Agriculture, Water:Yearbook of Agriculture, 1955 (GPO); J.J. Geraghty et al., Water Atlas of the United States (Port Washington NY: Water Information Center, 1973); U.S. Water Resources Council, The Nation's Water Resources: The First National Assessment (GPO, 1968); N. Wollman and G.W. Bonem, The Outlook for Water: Quality, Quantity, and National Growth (Johns Hopkins Univ. Press for Resources for the Future, 1971); G.F. White, ed., Strategies of American Water Management (Univ. of Michigan Press, rev.edn., 1971) and W.R.D. Sewell, "Water Across the American Continent," Geographical Magazine 46 (1974). The National Atlas includes many valuable maps of minerals and water.

Important references on climatology include R.A. Bryson and F.K. Hare, eds., Climates of North America (Elsevier, 1974); G.H.T. Kimble, Our American Weather (McGraw-Hill, 1955); S.S. Visher, Climatic Atlas of the United States (Harvard Univ. Press, 1954); Department of Agriculture, Climate and Man: Yearbook of Agriculture, 1941 (GPO), which includes a valuable series of contributions by geographers entitled "Climate and Agricultural Settlement"; publications of the U.S. Weather Bureau such as the series "Climates of the United States," reprinted in Climates of the States Based on National Oceanic and Atmospheric Administration Data, 2 vols. (Gale Research Co., 1978); statistical data on climate reported annually in the Statistical Abstract of the United States; climatic data in V. Showers, World Facts and Figures (Wiley, 1979); the many climatic maps in the National Atlas; American Meteorological Society, The History of American Weather, 4 vols. (Science History Publications, 1977); and G.R. Stewart's well-known novel, Storm (Modern Library, 1947).

We turn now to references on general ecology, biogeography, and soils. Works on general ecology include V.E. Shelford, The Ecology of North America (Univ. of Illinois Press, 1963), and, for general readers, P. Farb, The Face of North America: The Natural History of a Continent (Harper and Row, 1963); P. Farb, The Land and Wildlife of North America (LIFE Nature Library; Time-Life Books, 1964, 1966); The American Wilderness, 13 vols. (Time-Life Books, 1972-1974) richly illustrated with photographs in colour; and the superbly pictorial volumes on ecological subjects issued by the Sierra Club. See also R.G. Bailey Ecoregions of the United States (map and text; U.S. Forest Service, 1976). On plant geography see J.L. Vankat, The Natural Vegetation of North America

(Wiley, 1979); R. Daubenmire, Plant Geography: With Special Reference to North America (Academic, 1978); A.W. Küchler, Potential Natural Vegetation of the Conterminous United States (map and text, American Geographical Society, 1964; map reprinted in the National Atlas and available as a separate sheet from U.S. Geological Survey); J.W. Barrett, ed., Regional Silviculture of the United States (Wiley, 2nd edn., 1980); U.S. Department of Agriculture, Trees: Yearbook of Agriculture 1949 (GPO); the many publications of the U.S. Forest Service, such as E.L. Little, Jr., Atlas of United States Trees, 5 vols. (GPO, 1971-1978) and the data used in compiling the map "Major Forest Types" in the National Atlas; T.S. Elias, The Complete Trees of North America: Field Guide and Natural History (D. Van Nostrand, 1980); U.S. Department of Agriculture, Grass: Yearbook of Agriculture, 1948 (GPO); J.C. Malin, The Grassland of North America: Prolegomena to its History (Lawrence KS: The Author, rev. edn., 1956); and J.E. Weaver, North American Prairie (Lincoln NB: Johnson Publishing Co., 1954). For information on soils, see U.S. Department of Agriculture, Soil: Yearbook of Agriculture, 1957 (GPO) and, from the same source, Soils and Men: Yearbook of Agriculture, 1938 (GPO); also, H.D. Foth and J.W. Schafer, Soil Geography and Land Use (Wiley, 1980); P.J. Gersmehl, "Soil Taxonomy and Mapping," AAG Annals 67 (1977); D. Steila, The Geography of Soils (Prentice-Hall, 1976); and the vast array of publications by the U.S. Soil Conservation Service, including the soils map compiled for the National Atlas.

A very large literature exists in the broad area of resources, land use, natural hazards, environmental perception, and environmental conservation and management. Writings on the history of public lands occupy a prominent place. Much of this literature has been stimulated by repeated surges of public concern for deterioration of the natural environment- a cause very much in evidence during recent decades. Trying to select a brief list of references from the overwhelming mass of material is a daunting task. The titles that follow seem reasonably central, and their reference lists will point the way for further searches. K.A. Hammond, G. Macinko, and W.B. Fairchild, Sourcebook on the Environment: A Guide to the Literature (Univ. of Chicago Press, 1978) is world-wide in scope, but with a particularly intensive focus on the United States. See also J.W. Watson and T. O'Riordan, eds., The American Environment: Perceptions and Policies (Wiley, 1976); W.L. Graf, "Resources, the Environment and the American Experience," Journal of Geography 75 (1976); F.F. Darling and J.P. Milton, eds., Future Environments of North America (Natural History Press, 1966); I. Burton and R.W. Kates, eds., Readings in Resource Management and Conservation (Univ. of Chicago Press, 1965); G.H. Smith, ed., Conservation of Natural Resources (Wiley, 4th edn., 1971); R.L. Parson, Conserving American Resources (Prentice-Hall, 3rd edn., 1972); I. Burton, R.W.

Kates, and G.F. White, The Environment as Hazard (OUP, 1978); U.S. Department of Agriculture, Land: Yearbook of Agriculture, 1958 (GPO); R.H. Jackson, Land Use in America (Wiley, 1981); F.J. Marschner, Land Use and Its Patterns in the United States (U.S. Department of Agriculture, Handbook 153, GPO, 1959) and Major Land Uses in the United States (map, U.S. Department of Agriculture, GPO, 1950); S. Baker and H.W. Dill, Jr., The Look of Our Land- An Airphoto Atlas of the Rural United States, 5 vols. (U.S. Department of Agriculture, GPO, 1971-1972); M. Clawson, America's Land and Its Uses (Johns Hopkins Univ. Press for Resources for the Future, 1972) and M. Clawson, Man and Land in the United States (Univ. of Nebraska Press, 1964); W.R. Van Dersal, The American Land: Its History and Uses (OUP, 1943); B.G. Rosenkrantz and W.A. Koelsch, eds., American Habitat: A Historical Perspective (Free Press, 1973); R.M. Robbins, Our Landed Heritage: The Public Domain, 1776-1970 (Univ. of Nebraska Press, 2nd edn., 1976); J.M. Petulla, American Environmental History (Boyd and Fraser, 1977); and A. Runta, National Parks: The American Experience (Univ. of Nebraska Press, 1979).

Political, Cultural, and Social Geography

These three subfields of geography are so elastic in content and definition that they are hard to disentangle for bibliographic purposes. Hence the groupings of titles in this section are somewhat arbitrary. Particularly is this true of some ethnic studies that are grouped here under social geography but could have been placed under cultural or political geography. It should be noted that many titles having a specific regional context have been placed under regional headings at the end of the chapter.

Explicit works on the political geography of the United States are not nearly so numerous as those on cultural or social geography, though an immense cognate literature exists in the fields of history, politics, government, and journalism. S.D. Brunn, Geography and Politics in America (Harper and Row, 1974) is a topically-organized work that includes numerous bibliographical citations. The main topics (chapters) include Identification and Attitudes; Perception of Political Space; Political Territoriality; Political Elites and Decision-Making; Organization of Space; Boundaries and Interaction; Geography and the Law; Political Cultures; Electoral Geography; Government Programs; Politics and the Environment; and Political Geography of the Future. See also R.E. Norris and L.L. Haring Political Geography (Merrill, 1980), which includes several chapters on the United States. On politico-geographic origins see D. Whittlesey, Chap. 9, "The United States:The Origins of a Federal State," and Chap. 10, "The United States: Expansion and Consolidation," in W.G. East and A.E. Moodie, eds., The Changing World: Studies in Political Geography (World Book Co., 1956) and J.V. Fifer, "Unity by Inclusion: Core

Area and Federal Area at American Independence," Geographical Journal 142 (1976). Geographers have written fairly extensively on the international and internal political boundaries of the United States; most such studies antedate American Geography: Inventory and Prospect (1954; cited earlier), and many are included in a reference list accompanying the chapter on political geography in that volume. A miscellaneous selection of other titles by geographers includes K.D. Harries and S.D. Brunn, The Geography of Laws and Justice: Spatial Perspectives on the Criminal Justice System (Praeger, 1978); K.R. Cox, Conflict, Power and Politics in the City: A Geographic View (McGraw-Hill, 1973); J.S. Adams, "The Geography of Riots and Civil Disorders in the 1960's," Economic Geography 67 (1972); C. Akatiff, "The March on the Pentagon," AAG Annals 64 (1974); J. Swauger, "Regionalism in the 1976 Presidential Election," Geographical Review 70 (1980); J. O'Loughlin and D.A. Berg, "The Election of Black Mayors, 1969 and 1973," AAG Annals 67 (1977); J. O'Loughlin, "The Election of Black Mayors, 1977," AAG Annals 70 (1980); R.L. Morrill, "Ideal and Reality in Reapportionment," AAG Annals 63 (1973) and, Morrill, "Redistricting Revisited," AAG Annals 66 (1976); D.O. Bushman and W.R. Stanley, "State Senate Reapportionment in the Southeast," AAG Annals 61 (1971); H.R. Smith and J.F. Hart, "The American Tarriff Map," Geographical Review 45 (1955); A.K. Fitzsimmons, "Environmental Quality as a Theme in Federal Legislation," Geographical Review 70 (1980); R.J. Johnston, "Congressional Committees and Department Spending: The Political Influence on the Geography of Federal Expenditures in the United States," Institute of British Geographers, Transactions, N.S.4 (1979) and, Johnston, The Geography of Federal Spending in the United States of America (Research Studies Press/Wiley, 1980); E.H. Wohlenberg, "Public Assistance Effectiveness by States," AAG Annals 66 (1976); and I. Sutton, "Sovereign States and the Changing Definition of the Indian Reservation," Geographical Review 66 (1976).

A tiny sampling of titles from cognate fields includes J.B. Brebner, North Atlantic Triangle: The Interplay of Canada, the United States, and Great Britain (Yale Univ. Press, 1945); D.J. Eleazar, American Federalism: A View from the States (Crowell, 2nd edn., 1972) and, Eleazar, Cities of the Prairie: The Metropolitan Frontier and American Politics (Basic Books, 1970); R.G. Healy and J.S. Rosenberg, Land Use and the States (Johns Hopkins Univ. Press for Resources for the Future, Inc., 2nd edn., 1979); S.C. Patterson, "The Political Culture of American States," Journal of Politics 30 (1968); I. Sharkansky, Regionalism in American Politics (Bobbs-Merrill, 1970); M. Parenti, "Ethnic Politics and the Persistence of Ethnic Voting," American Political Science Review 61 (1967); K. Phillips, The Emerging Republican Majority (Doubleday-Anchor Books, 1970) and, Phillips, "The Balkanization of America," Harper's, May, 1978; N.R. Peirce, The Megastates of America

(Norton, 1972) and other works by Peirce listed under subsequent regional headings in this chapter; and K. Sale, _Power Shift: The Rise of the Southern Rim and its Challenge to the Eastern Establishment_ (Random House,1975). The general works by Brunn and by Norris and Haring, cited above, contain very extensive citations to titles in cognate fields.

Much useful material on the _cultural geography_ of the United States, including bibliography, is contained in general textbooks on cultural geography. See especially T.G. Jordan and L. Rowntree, _The Human Mosaic: A Thematic Introduction to Cultural Geography_ (Harper and Row, 2nd edn., 1979). This volume contains a large number of striking maps, extensive reference lists, and lengthy discussions of such topics as folk culture, popular culture, linguistic geography, and ethnic culture regions in the United States. The basic volume devoted exclusively to the United States is W. Zelinsky, _The Cultural Geography of the United States_ (Prentice-Hall, 1973). Though relatively short, it ranges over a wide spectrum of ideas and interpretations in provocative fashion, and it provides a valuable guide to bibliography For a useful summation in brief compass, see Bruce Bigelow, "Roots and Regions: A Summary Definition of the Cultural Geography of America," _Journal of Geography_ 79 (1980). This article includes an extensive set of reference footnotes and basic maps of cultural phenomena. North American Cultural Survey, _This Remarkable Continent_ (Texas A and M Univ. Press, 1982) is a wide-ranging cultural atlas. See also R.D. Gastil, _Cultural Regions of the United States_ (Univ. of Washington Press, 1975). Though not written by a geographer, this book makes much use of material by geographers. It includes a useful list of selected references. The same is true of G.S. Dunbar, "Illustrations of the American Earth: A Bibliographical Essay on the Cultural Geography of the United States," _American Studies: An International Newsletter_ 12 (1973). Vernacular, or popular, regions, "the product of the spatial perception of average people," are discussed in W. Zelinsky, "North America's Vernacular Regions," AAG _Annals_ 70 (1980). See also R.D. Mitchell, "The Formation of Early American Cultural Regions," in J.R. Gibson, ed., _European Settlement and Development in North America_, cited earlier.

On the American cultural landscape, see J.F. Hart, _The Look of the Land_ (Prentice-Hall, 1975); D.W. Meinig, ed., _The Interpretation of Ordinary Landscapes_ (OUP, 1979); National Research Council, Committee to Select Topographic Quadrangles Illustrating Cultural Geography, _Rural Settlement Patterns in the United States as Illustrated on One Hundred Topographic Quadrangle Maps_ (Washington, D.C., 1956); references accompanying the chapter "Settlement Geography" in _American Geography: Inventory and Prospect_, cited earlier; a series of thoughtful and engaging essays by D. Lowenthal, including "The American Scene," _Geographical Review_ 58 (1968), "The Bicentennial Landscape: A Mirror Held Up to the Past," Geographical Review 67 (1977), and "The Place of the Past in the American Landscape," in D. Lowenthal and M.J. Bowden, eds., Geographies of the Mind: Essays in Historical Geosophy in Honor of John Kirtland Wright (OUP, 1976); P.F. Lewis, D. Lowenthal, and Y-F. Tuan, Visual Blight in America (AAG Commission on College Geography, Resource Paper 23, 1973); F.B. Kniffen, "Folk Housing: Key to Diffusion," AAG Annals 55 (1965); P.F. Lewis, "Common Houses, Cultural Spoor," Landscape 19 (1975); E.T. Price,"The Central Courthouse Square in the American County Seat," AAG Annals 58 (1968); F.B. Kniffen," The American Agricultural Fair," AAG Annals 39 (1949) and 41 (1951); E.C. Mather and J.F. Hart, "Fences and Farms," Geographical Review 44 (1954), and, Hart and Mather, "The Character of Tobacco Barns and their Role in the Tobacco Economy of the United States," AAG Annals 51 (1961)

Other selected titles by geographers include K.B. Raitz, "Themes in the Cultural Geography of European Ethnic Groups in the United States," Geographical Review 69 (1979), and, Raitz, "Ethnic Maps of North America," Geographical Review 68 (1978); W. Zelinsky, "Selfward Bound? Personal Preference Patterns and the Changing Map of American Society," Economic Geography 50 (1974); J.R. Shortridge, "Patterns of Religion in the United States," Geographical Review 66 (1976);W. Zelinsky,"An Approach to the Religious Geography of the United States: Patterns of Church Membership in 1952," AAG Annals 51 (1961); J.W. Meyer, "Ethnicity, Theology, and Immigrant Church Expansion," Geographical Review 65 (1975); W.K. Crowley,"Old Order Amish Settlement: Diffusion and Growth," AAG Annals 68 (1978); W. Zelinsky, "Classical Town Names in the United States: The Historical Geography of an American Idea," Geographical Review 57 (1967) and, Zelinsky, "Cultural Variation in Personal Name Patterns in the Eastern United States," AAG Annals 60 (1970); J.F. Rooney, Jr., A Geography of American Sport (Addison-Wesley, 1974) and, Rooney, "Up from the Mines and Out from the Prairies: Some Geographical Implications of Football in the United States," Geographical Review 59 (1969); numerous articles in the journals Landscape (3 issues a year), Journal of Cultural Geography (semiannual), and Pioneer America: The Journal of Historic American Material Culture (quarterly); and G.F. Carter, Earlier Than You Think: A Personal View of Man in America (Texas A&M Univ. Press, 1980).

In cognate fields see M. Lerner, America as a Civilization: Life and Thought in the United States Today (Simon and Schuster, 1957); G.R. Stewart, Names on the Land: A Historical Account of Place-Naming in the United States (Houghton Mifflin, 1945), and Stewart, American Ways of Life (Doubleday, 1954; reissued by Russell and Russell, 1971); W.N. Morgan, Prehistoric Architecture in the Eastern United States (MIT Press, 1980); H. Glassie, Pattern in the Folk Culture of the Eastern United States (Univ. of Pennsylvania Press, 1968); C.A. Weslager, The Log Cabin in America from Pioneer Days to the

Present (Rutgers Univ. Press, 1969); E. Arthur and D. Witney, The Barn: A Vanishing Landscape in North America (McClelland and Stewart, 1972); H.M. Jones, O Strange New World - American Culture: The Formative Years (Viking, 1964) and Jones, The Age of Energy - Varieties of American Experience, 1865-1915 (Viking, 1971); E.S. Gaustad, Historical Atlas of Religion in America (Harper and Row, rev. edn., 1976); N.D. Glenn and J.L. Simmons, "Are Regional Cultural Differences Diminishing?" Public Opinion Quarterly 31 (1967); N.D. Glenn and L. Hill, Jr., "Rural-Urban Differences in Attitudes and Behavior in the United States," Annals of the American Academy of Political and Social Science 429 (1977); and J.A. Hague, ed., American Character and Culture in a Changing World (Greenwood, 1979).

Social geography, including many aspects of ethnicity, has been the subject of numerous books and articles during recent years. We may begin with J.W. Watson, Social Geography of the United States (Longmans,1979), which, like most recent works in this field, centres attention on important social problems. It cites a varied list of references in chapter bibliographies. See also the material on the United States in J.A. Jakle et al., Human Spatial Behavior: A Social Geography (Duxbury, 1976). Works by geographers centring on wealth, poverty, and quality of life include D.M. Smith, The Geography of Social Well-Being in the United States: An Introduction to Territorial Social Indicators (McGraw-Hill, 1973); R.L. Morrill and E.H. Wohlenberg, The Geography of Poverty in the United States (McGraw-Hill, 1973); R. Peet and R. Elgie, "Bibliography on American Poverty," Antipode 2 (1970); R. Peet, ed., Geographical Perspectives on American Poverty (Antipode, Worcester MA, Monographs in Social Geography 1, 1972); and A.F. Burghardt, "Income Density in the United States," AAG Annals 62 (1972). On population, see "The Population of Anglo-America" in R.P. Larkin, G.L. Peters, and C.H. Exline, People, Environment, and Place: An Introduction to Human Geography (Merrill, 1981); the material on population in the general geographies of the United States and Canada cited earlier; W. Zelinsky, "Changes in the Geographic Patterns of Rural Population in the United States, 1790-1960," Geographical Review 52 (1962); S.M. Golant, Location and Environment of Elderly Population (Wiley, 1979); D.S. Massey, "Residential Segregation and Spatial Distribution of a Non-Labor Force Population: The Needy Elderly and Disabled," Economic Geography 56 (1980); J.S. Fisher and R.L. Mitchelson, "Forces of Change in the American Settlement Pattern," Geographical Review 71 (1981); G.V. Fugitt, P.R. Voss, and J.C. Doherty, Growth and Change in Rural America (Washington, D.C., Urban Land Institute, 1979); T.O. Graff and R.F. Wiseman, "Changing Concentrations of Older Americans," Geographical Review 68 (1978); J.P. Allen, "Changes in the American Propensity to Migrate," AAG Annals 67 (1977); and C.C. Roseman, Changing Migration Patterns within the United States (AAG Resource Papers for College Geography 77-2 (1977).

On ethnicity and ethnic problems, see J.A. Jakle, Ethnic and Racial Minorities in North America: A Selected Bibliography of the Geographical Literature (Council of Planning Librarians, 1973); G.A. Davis and O.F. Donaldson, Blacks in the United States: A Geographic Perspective (Houghton Mifflin, 1975); R. Ernst and L. Hugg, Black America: Geographic Perspectives (Doubleday-Anchor Books, 1976); H.M. Rose, ed., Contributions to an Understanding of Black America, special issue, Economic Geography 48 (1972); W.C. Calef and H.J. Nelson, "Distribution of Negro Population in the United States," Geographical Review 46 (1956); J.F. Hart, "The Changing Distribution of the American Negro," AAG Annals 50 (1960); H.M. Rose, The Black Ghetto: A Spatial Behavioral Perspective (McGraw-Hill, 1971); R.L. Morrill, "The Negro Ghetto: Problems and Alternatives," Geographical Revue 55 (1965) E.M. Neils, Reservation to City: Indian Migration and Federal Relocation (Univ. of Chicago, Geography Research Paper 131, 1971); D.J. Wishart, "The Dispossession of the Pawnee," AAG Annals 69 (1979); T.D. Boswell and T.C. Jones, "A Regionalization of Mexican Americans in the United States," Geographical Review 70 (1980); and R.L. Nostrand, "The Hispanic-American Borderland: Delimitation of an American Culture Region," AAG Annals 60 (1970). On the geography of crime see D.E. Georges-Abbeyie and K.D. Harries, Crime: A Spatial Perspective (Columbia Univ. Press, 1980); K.D. Harries, The Geography of Crime and Justice (McGraw-Hill, 1974); and D.E. Georges, The Geography of Crime and Violence: A Spatial and Ecological Perspective (AAG Resource Papers for College Geography 78-1, 1978).

Selected references on social geography, from government sources and in cognate fields include the publications of the U.S. Bureau of the Census such as Current Population Reports and the elaborate series of large population maps prepared by the Geography Division; G. Sternlieb and J.W. Hughes, Current Population Trends in the United States (Center for Urban Policy Research, Rutgers Univ., 1978); D.L. Brown and J.M. Wardwell, eds., New Directions in Urban-Rural Migration: The Population Turnaround in Rural America (Academic, 1980); B. Weinstein and R. Firestine, Regional Growth and Decline in the United States: The Rise of the Sunbelt and the Decline of the Northeast (Praeger, 1978); various publications of the Population Reference Bureau, Washington, D.C., such as The United States Population Data Sheet (1981); A. Cooke, The Americans (Knopf, 1979); B.A. Weisberger and the Editors of American Heritage, The American Heritage History of the American People (American Heritage Publishing Co., 1970, 1971); J.C Furnass, The Americans: A Social History of the United States, 1587-1914 (Putnam, 1969); R.A. Bartlett, The New Country: A Social History of the American Frontier, 1776-1890 (OUP, 1974); R. Berthoff, An Unsettled People: Social Order and Disorder in American History (Harper and Row, 1971); M.L. Hansen, The Immigrant in American History (Harvard Univ. Press,

(1940); M.A. Jones, American Immigration (Univ. of Chicago Press, 1960); M.M. Gordon, Assimilation in American Life: The Role of Race, Religion, and National Origin (OUP, 1964); S. Thernstrom, ed., Harvard Encyclopedia of American Ethnic Groups (Harvard Univ. Press, 1980); D.K. Fellows, A Mosaic of America's Ethnic Minorities (Wiley, 1972); D.M. Johnson and R.R. Campbell, Black Migration in America : A Social Demographic History (Duke Univ. Press, 1981); H.E. Driver, Indians of North America (Univ. of Chicago Press, 2nd rev. edn., 1969); W.E. Washburn, The Indian in America (Harper and Row, 1975); A.M. Josephy, Jr., The Indian Heritage of America (Knopf, 1969); E.E. Edwards, "American Indian Contributions to Civilization," Minnesota History 15 (1934); A.L. Sorkin, The Urban American Indian (Heath, 1979); D.J. Weber, ed., Foreigners in their Native Land: Historical Roots of the Mexican Americans (Univ. of New Mexico Press, 1975); S. Weintraub and S.R. Ross, The Illegal Alien from Mexico (Univ. of Texas Press, 1980); and J. Chen, The Chinese of America: from the Beginnings to the Present (Harper and Row, 1980). See also such journals as Growth and Change, American Sociological Review, American Journal of Sociology, and Social Forces.

Economic Geography

At present there seems to be no single publication providing an adequate full-scale treatment of the economic geography of the United States. However, the persistent researcher can build up a reasonably comprehensive picture from the general geographies of the United States and Canada cited on pp.131-135 or from general textbooks on economic or human geography. A selection of the latter includes J.O. Wheeler and P.O. Muller, Economic Geography (Wiley, 1981); J.H. Butler, Economic Geography: Spatial and Environmental Aspects of Economic Activity (Wiley, 1980); T.J. Wilbanks, Location and Well-Being: An Introduction to Economic Geography (Harper and Row, 1980); J.W. Alexander and L.J. Gibson, Economic Geography (Prentice-Hall, 2nd edn., 1979); R.L. Morrill and J.M. Dormitzer, The Spatial Order: An Introduction to Modern Geography (Duxbury, 1979); R.R. Boyce, The Bases of Economic Geography (Holt, Rinehart and Winston, 2nd edn., 1978); P.E. Lloyd and P. Dicken, Location in Space: A Theoretical Approach to Economic Geography (Harper and Row, 2nd edn., 1977); B.J.L. Berry, E.C. Conkling, and D.M. Ray, The Geography of Economic Systems (Prentice-Hall, 1976); E.C. Conkling and M. Yeates, Man's Economic Environment (McGraw-Hill, 1976); R.S. Thoman and P.B. Corbin, The Geography of Economic Activity (McGraw-Hill, 3rd edn., 1974); and J.F. Kolars and J.D. Nystuen, Human Geography: Spatial Design in World Society (McGraw-Hill, 1974). From the materials on the United States in the foregoing sources, one could assemble a large and diversified compendium of text, maps, graphs, charts, diagrams, tables, photographs, and bibliography. Cartographic information in profusion also is available from such sources as J.D. Chapman and J.C. Sherman, eds., Oxford Regional Economic Atlas: The United States and Canada (OUP, 2nd edn., 1975); The National Atlas of the United States of America; Goode's World Atlas and other general atlases; and the cartographic publications of the U.S. Bureau of the Census. Some recent professional articles of broad import written by geographers include E.J. Malecki, "Government-Funded R&D: Some Regional Economic Implications," Professional Geographer 33 (1981); J.E. McConnell, "Foreign Direct Investment in the United States," AAG Annals 70 (1980); W.B. Beyers, "Contemporary Trends in the Regional Economic Development of the United States", Professional Geographer 31 (1979); C.E. Browning and W. Gesler, "The Sun Belt-Snow Belt: A Case of Sloppy Regionalizing," Professional Geographer 31 (1979); J.R. Borchert, "Major Control Points in American Economic Geography," AAG Annals 68 (1978); R. Estall, "Regional Planning in the United States: An Evaluation of Experience under the 1965 Economic Development Act," Town Planning Review 48 (1977); and various specialized, technical, and theoretical articles in the journal Economic Geography. The reader may wish to consult the latter journal's book reviews and lists of books received. See also the collected articles in E.L. Ullman, Geography as Spatial Interaction, ed. R.R. Boyce (Univ. of Washington Press, 1978), as well as B.J.L. Berry, "The Geography of the United States in the Year 2000," IBG Transactions 51 (1970), and W. Goodwin, "The Management Center in the United States," Geographical Review 55 (1965).

Space permits only a few citations of a general character from the huge literature of such cognate fields as economic history, planning, and regional science. Two recent works of high importance are G. Porter, ed., Encyclopedia of American Economic History: Studies of the Principal Movements and Ideas, 3 vols. (Scribner's, 1980), and W.K. Hutchinson, ed., American Economic History: A Guide to Information Sources (Gale Research Co., 1980). See also S. Salsbury, "Economic History Then and Now: 'The Economic History of the United States' in the Light of Recent Scholarship," Agricultural History 53 (1979), a review of an eight-volume work; E.J. Perkins, The Economy of Colonial America (Columbia Univ. Press, 1980); G.M. Walton and J.F. Shepherd, The Economic Rise of Early America (Cambridge Univ.Press, 1979); R.M. Robertson, History of the American Economy (Harcourt, 3rd edn., 1979); S.P. Lee and P. Passell, A New Economic View of American History (Norton, 1979); W. W. Rostow, Getting from Here to There: America's Future in the World Economy (McGraw-Hill, 1978); B.L. Weinstein and R.E. Firestine, Regional Growth and Decline in the United States: The Rise of the Sunbelt and the Decline of the Northeast (Praeger, 1978); J. Rivkin and R. Barber, The North Will Rise Again: Pensions, Politics, and Power in the 1980's (Beacon

Press, 1978); G. Sternlieb and J.W. Hughes, eds., Post-Industrial America: Metropolitan Decline and Inter-Regional Job Shifts (Center for Urban Policy Research, 1975); A. Groner and the Editors of American Heritage and Business Week, The American Heritage History of American Business and Industry (American Heritage, 1972); J.H. Cumberland, Regional Development: Experiences and Prospects in the United States of America (Mouton, 1971); D.J. Bogue and C.L. Beale, Economic Areas of the United States (Free Press, 1961); and H.S. Perloff et al., Regions, Resources, and Economic Growth (Johns Hopkins Univ. Press, 1960). Enormous amounts of information on the American economy pour from the presses; a selection includes such scholarly quarterlies as the Journal of Economic History, Growth and Change, and the Journal of Regional Science; various volumes of papers and proceedings issued by the Regional Science Association; journals in the field of planning; and such mass-circulation media as Fortune, Business Week, World Business Weekly, and general news magazines, the New York Times, and the Wall Street Journal.

We turn now to sources of information in subfields of economic geography. For many purposes the general references already cited may prove sufficient, but a large volume of more specialized literature does exist on the geography of agriculture, manufacturing, and other branches of economic activity. Attention will be focussed on references applicable to the United States as a whole, with those of a more localized cast being reserved for the regional discussions that conclude the chapter.

Numerous references to the agricultural geography of the United States are contained in such general works by geographers as H.F. Gregor, Geography of Agriculture: Themes in Research (Prentice-Hall, 1970) and D.B. Grigg, The Agricultural Systems of the World: An Evolutionary Approach (Cambridge Univ. Press, 1974). More specifically on the United States, see E.G. Smith, Jr., "America's Richest Farms and Ranches," AAG Annals 70 (1980); T.L. McKnight, "Great Circles on the Great Plains: The Changing Geometry of American Agriculture," Erdkunde 33 (1979); P.J. Gersmehl, "No-Till Farming: The Regional Application of a Revolutionary Agricultural Technology," Geographical Review 68 (1978); R.H. Platt, "The Loss of Farmland: Evolution of Public Policy," Geographical Review 67 (1977); A. Chominot, "Quelques tendances nouvelles de l'agriculture des U.S.A.", Annales de Géographie 467 (1976); E.G. Smith, "Fragmented Farms in the United States," AAG Annals 65 (1975); S.R. Jumper, "Wholesale Marketing of Fresh Vegetables," AAG Annals 64 (1974); P.O. Muller, "Trend Surfaces of American Agricultural Patterns: A Macro-Thünian Analysis," Economic Geography 49 (1973); H.F. Gregor, "The Large Industrialized American Crop Farm: A Mid-Latitude Plantation Variant," Geographical Review 60 (1970) and Gregor, "Farm Structure in Regional Comparison : California and New Jersey,"

Economic Geography 45 (1969); J.F. Hart, "A Map of the Agricultural Implosion," AAG Proceedings 2 (1970), also, Hart, "Loss and Abandonment of Cleared Farm Land in the Eastern United States," AAG Annals 58 (1968) and "The Changing American Countryside," in S.B. Cohen, ed., Problems and Trends in American Geography (Basic Books, 1967); J.R. Peet, "The Spatial Expansion of Commercial Agriculture in the United States in the Nineteenth Century: A von Thünen Interpretation," Economic Geography 45 (1969); E. Higbee, American Agriculture: Geography, Resources, Conservation (Wiley, 1958); C.D. Harris, "Agricultural Production in the United States: The Past Fifty Years and the Next," Geographical Review 47 (1957); and, as a classic though outdated study, O.E. Baker, "Agricultural Regions of North America," Economic Geography 2 (1926). A few cognate works of major import include L. Soth, "The Grain Export Boom: Should It Be Tamed?", Foreign Affairs 59 (1981); W.W. Cochrane, The Development of American Agriculture: A Historical Analysis (Univ. of Minnesota Press, 1979); W. Ebeling, The Fruited Plain: The Story of American Agriculture (Univ. of California Press, 1979); R.M. Fraenkel et al., The Role of U.S. Agriculture in Foreign Policy (Praeger, 1979); U.S. Senate, 99th Congress, The State of American Agriculture: Hearings before the Committee on Agriculture, Nutrition, and Forestry, March 6-10, 1978 (GPO, 1978); T.R. Ford, ed., Rural U.S.A.: Persistence and Change (Iowa State Univ. Press, 1978); E. Heady, "The Agriculture of the United States," Scientific American 235 (1976); J.T. Schlebecker, Whereby We Thrive: A History of American Farming, 1607-1972 (Iowa State Univ. Press, 1975); U.S. Soil Conservation Service, map, Land Resource Regions and Major Land Resource Areas of the United States (1963); J.T. Schlebecker, "The World Metropolis and the History of American Agriculture," Journal of Economic History 20 (1960); and L. Haystead and G.C. Fite, The Agricultural Regions of the United States (Univ. of Oklahoma Press, 1955). Also of fundamental importance are reports and maps issued by the U.S. Bureau of the Census, particularly the Graphic Summary of the Census of Agriculture every five years, and such maps as Crop Patterns in the United States: 1959 and Livestock and Livestock Products Sold in the United States: 1959, both drawn by G.F. Jenks; the vast array of publications of the U.S. Department of Agriculture, notably Agricultural Statistics (annual), the long series of Yearbooks of Agriculture, and the map Major Land Uses of the United States by F.J. Marschner (1950); the articles, reviews, and booklists in the quarterly journal Agricultural History issued by the Agricultural History Society and the bibliographies issued by the Agricultural History Center, Univ. of California, Davis CA. Information on forestry and fisheries is scattered through innumerable references cited in this chapter. On forestry, also see the publications of the U.S. Forest Service; the articles in the monthly magazine American Forests; and R.G. Lillard,

The Great Forest (Knopf, 1947).

On industrial geography (manufacturing and mining), selected titles by geographers include F.C.F. Earney, "The Geopolitics of Minerals," Focus 31 (1981); J.E. McConnell, "Foreign Ownership and Trade of United States High-Technology Manufacturing," Professional Geographer 33 (1981); I.M. Sheskin, "A Geographic Approach to the Study of Natural Gas," Journal of Geography 79 (1980); M.D. Winsberg "The Changing Concentration of U.S. Industrial Employment, 1940-1977," Journal of Geography 79 (1980); R.E. Lonsdale and H.L. Seyler, eds., Nonmetropolitan Industrialization (Wiley, 1979); J. Rees, "Technological Change and Regional Shifts in American Manufacturing," Professional Geographer 31 (1979); D.W. Clements, "Recent Trends in the Geography of Coal," AAG Annals 67 (1977); E.W. Miller, Manufacturing: A Study of Industrial Location (Pennsylvania State Univ. Press, 1977); J.P. Osleeb and I.M. Sheskin, "Natural Gas: A Geographical Perspective," Geographical Review 67 (1977); B.W. Smith, "Analysis of the Location of Coal-Fired Power Plants in the Eastern United States," Economic Geography 49 (1973); K. Warren, The American Steel Industry, 1850-1970: A Geographical Interpretation (OUP, 1973); L.J. Gibson, "An Analysis of the Location of Instrument Manufacture in the United States," AAG Annals 60 (1970); F.C.F. Earney, "New Ores for Old Furnaces: Pelletized Iron," AAG Annals 59 (1969); P.F. Mason, "Some Changes in Domestic Iron Mining as a Result of Pelletization," AAG Annals 59 (1969); J.L. Morrison, M.W. Scripter, and R.H.T. Smith, "Basic Measures of Manufacturing in the United States, 1958," Economic Geography 44 (1968); G. Alexandersson, Geography of Manufacturing (Prentice-Hall, 1967); A.R. Pred, "The Intrametropolitan Location of American Manufacturing," AAG Annals 54 (1964); W. Zelinsky, "Has American Industry Been Decentralizing? The Evidence for the 1939-1954 Period," Economic Geography 38 (1962); G. Alexandersson, The Industrial Structure of American Cities (Univ. of Nebraska Press, 1956); and C.D. Harris, "The Market as a Factor in the Localization of Industry in the United States," AAG Annals 44 (1954). An abbreviated list of cognate references includes D.J. Cuff and W.J. Young, The United States Energy Atlas (Free Press/MacMillan, 1980); L. Kiers, The American Steel Industry: Problems, Challenges, Perspectives (Westview, 1980); U.S. National Academy of Sciences, Energy in Transition, 1985-2010 (Freeman, 1980); U. Lantzke, "Expanding World Use of Coal," Foreign Affairs 58 (1979-80); E.L. Allen, Energy and Economic Growth in the United States (MIT Press, 1979); A.E. Eckes, Jr., The United States and the Global Struggle for Minerals (Univ. of Texas Press, 1979); L.C. Hunter, A History of Industrial Power in the United States, 1780-1930, Vol. 1, Waterpower in the Century of the Steam Engine (Univ. Press of Virginia for Eleutherian Mills-Hagley Foundation, 1979); R.D. Norton and J. Rees, "The Product Cycle and the Spatial Decentralization of American Manufacturing," Regional Studies 13 (1979); S.H. Schurr et al., Energy in America's Future (Johns Hopkins Univ. Press for Resources for the Future, 1979); R.M. Christie, "The City in the American Industrial System," Growth and Change 7 (1976); R.L. Nelson, Concentration in the Manufacturing Industries of the United States: A Mid-Century Report (Yale Univ. Press, 1963); V.R. Fuchs, Changes in the Location of Manufacturing in the United States since 1929 (Yale Univ. Press, 1962); and V.S. Clark, History of Manufactures in the United States, 3 vols. (Carnegie Institution of Washington, 1929; reprinted by Peter Smith, 1949).

Among the relatively small number of titles by geographers on transportation, trade, and other service types of activity are "Transport in the USA," general ed. J.F. Davis, a series of articles in consecutive issues of the Geographical Magazine 53 (1981); P.W. Daniels, ed., Spatial Patterns of Office Growth and Location (Wiley, 1979); E.W. Richards and M.L. Thaller," United States Railway Traffic: An Update," Professional Geographer 30 (1978); E. Schenker, H.M. Mayer, and H.C. Brockel, The Great Lakes Transportation System (Univ. of Wisconsin Press, 1976); J.T. Starr, Jr., The Evolution of the Unit Train, 1960-1969 (Univ. of Chicago, Geography Research Paper 158, 1976); M.E.E. Hurst, ed., Transportation Geography: Comments and Readings (McGraw-Hill, 1974); G.W. Shannon and G.E.A. Dever, Health Care Delivery: Spatial Perspectives (McGraw-Hill, 1974); W.F. Wacht, The Domestic Air Transportation Network of the United States (Univ. of Chicago, Geography Research Paper 154, 1974); H.M. Mayer, "Some Geographic Aspects of Technological Change in Maritime Transportation," Economic Geography 49 (1973); E.J. Taaffe and H.L. Gauthier, Jr., Geography of Transportation (Prentice-Hall, 1973); J.B. Kenyon, "Elements in Interport Competition in the United States," Economic Geography 46 (1970); J.E. Vance, Jr., The Merchant's World: The Geography of Wholesaling (Prentice-Hall, 1970); B.J.L. Berry, Geography of Market Centers and Retail Distribution (Prentice-Hall, 1967); R.S. Thoman and E.C. Conkling, Geography of International Trade (Prentice-Hall, 1967); C.C. Colby, North Atlantic Arena: Water Transport in the World Order (Southern Illinois Univ. Press, 1966); G. Alexandersson and G. Norström, World Shipping: An Economic Geography of Ports and Seaborne Trade (Wiley, 1963); W.H. Wallace, "Freight Traffic Functions of Anglo-American Railroads," AAG Annals 53 (1963); W.L. Garrison et al., Studies of Highway Development and Geographic Change (Univ. of Washington Press, 1959); W.H. Wallace, "Railroad Traffic Densities Patterns," AAG Annals 48 (1958); E.L. Ullman, American Commodity Flow (Univ. of Washington Press, 1957); D. Patton, "The Traffic Pattern on American Inland Waterways," Economic Geography 32 (1956); and E.J. Taaffe, "Air Transportation and United States Urban Distribution," Geographical Review 46 (1956). A few cognate references include N.K. Taneja,

*The U.S. Airfreight Industry* (Lexington Books, 1979); R.T. Green and J.M. Lutz, *The United States and World Trade* (Praeger, 1978); and P.H. Cootner, "The Role of the Railroads in United States Economic Growth," *Journal of Economic History* 23 (1963). Finally, one cannot overemphasize the utility of U.S. government statistical, cartographic, and other publications for all fields of economic geography. Two examples in the field of transportation are the elaborately detailed statistics in Corps of Engineers, U.S. Army, *Waterborne Commerce of the United States* (annual), and the set of 19 maps in the series *National Energy Transportation Systems*, published for the Committees on Commerce and Interior and Insular Affairs of the U.S. Senate by the U.S. Geological Survey in 1976.

*Urban Geography*

American geographers have assiduously tilled the field of urban geography in recent decades. The results of this work are summarized and explicated in a series of recent textbooks on American cities, or in more general texts on urban geography that deal extensively with American cities. Titles include R. Palm, *The Geography of American Cities* (OUP, 1981); T.A. Hartshorn, *Interpreting the City: An Urban Geography* (Wiley, 1980); M. Yeates and B.J. Garner, *The North American City* (Harper and Row, 3rd edn., 1980); R.M. Northam, *Urban Geography* (Wiley, 2nd edn., 1979); D.S. Rugg, *Spatial Foundations of Urbanism* (Brown, 2nd edn., 1979); J.E. Vance, Jr., *This Scene of Man: The Role and Structure of the City in the Geography of Western Civilization* (Harper and Row, 1977); and R.E. Murphy, *The American City: An Urban Geography* (McGraw-Hill, 2nd edn., 1974). These books are extensively provided with graphics and bibliography. Much useful material also may be found in H.M. Mayer and C.F. Kohn, eds., *Readings in Urban Geography* (Univ. of Chicago Press, 1959), and in R.D. Swartz *et al.*, *Metropolitan America: Geographic Perspectives and Teaching Strategies* (National Council for Geographic Education, 1972). A valuable historical overview is offered in P.G. Goheen, "Interpreting the American City: Some Historical Perspectives," *Geographical Review* 64 (1974). For studies of the largest cities, see R. Abler and J.S. Adams, *A Comparative Atlas of America's Great Cities: Twenty Metropolitan Regions* (Association of American Geographers and Univ. of Minnesota Press, 1976) and J.S. Adams, ed., *Contemporary Metropolitan America*, 4 vols. (Ballinger, 1976). The latter work comprises twenty geographical vignettes of individual metropolises, together with a general introduction by J.E. Vance, Jr. entitled "The American City: Workshop for a National Culture." Titles of specific vignettes are cited under appropriate regions in the last part of this chapter; thirteen are available from the publisher as separate monographs. For a review of the metropolitan atlas and the entire series of vignettes, see J. Gottmann, "The Mutation of the American City: A Review of the Comparative Metropolitan Analysis Project," Geographical Review 68 (1978). Classic studies of the structure and classification of cities include C.C. Colby, "Centrifugal and Centripetal Forces in Urban Geography," AAG Annals 23; (1933); C. Harris and E.L. Ullman, "The Nature of Cities," Annals of the American Academy of Political and Social Science 242 (November, 1945); C.D. Harris, "A Functional Classification of Cities in the United States," Geographical Review 33 (1943); and H.J. Nelson, "A Service Classification of American Cities," Economic Geography 31 (1955). All four of the foregoing selections are reprinted in Mayer and Kohn, op. cit. See also R.H.T. Smith, "Method and Purpose in Functional Town Classification," AAG Annals 55 (1965), a review article.

Selected historical studies of cities by geographers include A. Pred, Urban Growth and City Systems in the United States, 1840-1860 (Harvard Univ. Press, 1980); M.P. Conzen, "The Maturing Urban System in the United States, 1840-1910," AAG Annals 67 (1977); C.V. Earle, "The First English Towns in North America," Geographical Review 67 (1977); J.E. Vance, Jr., "Cities in the Shaping of the American Nation," Journal of Geography 75 (1976); D. Ward, Cities and Immigrants: A Geography of Change in Nineteenth-Century America (OUP, 1971); J.R. Borchert, "American Metropolitan Evolution," Geographical Review 57 (1967); A.R. Pred, The Spatial Dynamics of U.S. Urban-Industrial Growth, 1800-1914 (MIT Press, 1966) and Pred, "Industrialization, Initial Advantage, and American Metropolitan Growth," Geographical Review 55 (1965).

A miscellany of other titles on U.S. urban geography by geographers includes P.O. Muller, Contemporary Suburban America (Prentice-Hall, 1981); B.J.L. Berry, "Inner City Futures: An American Dilemma Revisited," IBG Transactions N.S.5 (1980); S.D. Brunn and J.O. Wheeler, eds., The American Metropolitan System: Present and Future (Wiley, 1980); M. Yeates, North American Urban Patterns (Wiley, 1980); L.R. Ford, "Urban Preservation and the Geography of the City in the USA," Progress in Human Geography 3 (1979); the chapters on urban geography in P. Haggett, Geography: A Modern Synthesis (Harper and Row, 3rd edn., 1979); D.M. Smith and D.T. Herbert, eds., Social Problems in the City (OUP, 1979); D.D. Baumann and D. Dworkin, Water Resources for Our Cities (AAG Resource Papers for College Geography 78-2, 1978); L.J. King and R.G. Golledge, Cities, Space and Behavior (Prentice-Hall, 1978); P.D. Phillips, "New Patterns of American Metropolitan Population Growth in the 1970's," Journal of Geography 77 (1978) and Phillips, "Slow Growth: A New Epoch of American Metropolitan Evolution," Geographical Review 68 (1978); B.J.L. Berry and Q. Gillard, The Changing Shape of Metropolitan America: Commuting Patterns, Urban Fields and Decentralization Processes, 1960-1970 (Ballinger, 1977); B.J.L. Berry and J.D. Kasarda, Contemporary Urban Ecology (Macmillan, 1977); K.R. Cox, ed., Urbanization and Conflict in Market Societies

(Maaroufa Press, 1977); J.F. Hart, "Urban Encroachment on Rural Areas," Geographical Review 67 (1977); T.R. Lakshmanan and L.R. Chatterjee, Urbanization and Environmental Quality (AAG Resource Papers for College Geography 77-1, 1977); R. Peet, Radical Geography: Alternative Viewpoints on Contemporary Social Issues (Methuen, 1977); A.R. Pred, City-Systems in Advanced Economies: Past Growth, Present Processes, and Future Development Options (Hutchinson, 1977); J.S. Adams, ed., Urban Policymaking and Metropolitan Dynamics (Ballinger, 1976); G.W. Carey, "Land Tenure, Speculation, and the State of the Aging Metropolis," Geographical Review 66 (1976); J.K. Mitchell, "Adjustment to New Physical Environments beyond the Metropolitan Fringe," Geographical Review 66 (1976); P.D. Phillips, "The Changing Standard Metropolitan Statistical Area," Journal of Geography 75 (1976); S. Gale and E. Moore, eds., The Manipulated City: Perspectives on Spatial Structure and Social Issues in Urban America (Maaroufa Press, 1975); A.M. Guest, "Population Suburbanization in American Metropolitan Areas," Geographical Analysis 7 (1975); D.W. Harvey, "The Political Economy of Urbanization in Advanced Capitalist Countries: The Case of the United States," in S. Gappert and H. Rose, eds., The Social Economy of Cities (Beverly Hills CA: Sage Publications, Urban Affairs Annual Reviews 9, 1975); P.O. Muller, The Outer City: Geographical Consequences of the Urbanization of the Suburbs (AAG Resource Papers for College Geography 75-2, 1975); A.R. Pred, "Diffusion, Organizational Spatial Structure, and City-System Development," Economic Geography 51 (1975); J. Sommer, "Fat City and Hedonopolis: The American Urban Future?" in R. Abler et al., Human Geography in a Shrinking World (Duxbury, 1975); G. Manners, "The Office in Metropolis: An Opportunity for Shaping Metropolitan America," Economic Geography 50 (1974); A.R. Pred, Major Job-Providing Organizations and Systems of Cities (AAG Commission on College Geography, Resource Paper 27, 1974); H.B Rodgers, "A Profile of the American City," in D. Welland, ed., The United States: A Companion to American Studies (Methuen, 1974); J.O. Wheeler, The Urban Circulation Noose (Duxbury, 1974); B.J.L. Berry, Growth Centers in the American Urban System, 2 vols. (Ballinger, 1973) and, Berry, The Human Consequences of Urbanization (St. Martin's, 1973); K.R. Cox, Conflict, Power and Politics in the City: A Geographic View (McGraw-Hill, 1973); T.A. Hartshorn, "Industrial/Office Parks: A New Look for the City," Journal of Geography 72 (1973); D.W. Harvey, Social Justice and the City (Johns Hopkins Univ. Press, 1973); F.E. Horton, ed., Geographical Perspectives and Urban Problems (Washington, D.C.: National Academy of Sciences, 1973); J.R. Borchert, "America's Changing Metropolitan Regions," in J.F. Hart, ed., Regions of the United States, op. cit., (1972); D.W. Harvey, Society, the City, and the Space-Economy of Urbanism (AAG Commission on College Geography, Resource Paper 18, 1972); R.E. Murphy, The Central

Business District (Aldine-Atherton, 1972); T.A. Hartshorn, "Inner City Residential Structure and Decline," AAG Annals 61 (1971); B.J.L. Berry and F.E. Horton, Geographic Perspectives on Urban Systems (Prentice-Hall, 1970); H.M. Mayer et al., A Modern City- Its Geography (National Council for Geographic Education, 1970; a reprint of nine articles from the Journal of Geography 68, 1969); H.M. Mayer, The Spatial Expression of Urban Growth (AAG Commission on College Geography, Resource Paper 7, 1969); H.M. Rose, Social Processes in the City: Race and Urban Residential Choice (AAG Commission on College Geography, Resource Paper 6, 1969); J. Gottmann and R.A. Harper, eds., Metropolis on the Move: Geographers Look at Urban Sprawl (Wiley, 1967); R.M. Northam, "Declining Urban Centers in the United States: 1940-1960," AAG Annals 53 (1963); and C.D. Harris, "The Pressure of Residential-Industrial Land Use," in W.L. Thomas, ed., Man's Role in Changing the Face of the Earth (Univ. of Chicago Press, 1956).

It is possible to list here only a few titles from the enormous literature on urbanism in fields cognate to geography. Among the numerous histories of American urban life and planning are H.P. Chudacoff, The Evolution of American Urban Society (Prentice-Hall, 2nd edn., 1981); J.C. Teaford, City and Suburb: The Political Fragmentation of Metropolitan America, 1850-1970 (Johns Hopkins Univ. Press, 1979); C.N. Glaab and A.T. Brown, A History of Urban America (Macmillan, 2nd edn., 1976); Z. Miller, The Urbanization of Modern America: A Brief History (Harcourt, 1973); S.B. Warner, Jr., The Urban Wilderness: A History of the American City (Harper and Row, 1972); J.W. Reps, The Making of Urban America: A History of City Planning in the United States (Princeton Univ. Press, 1965); and portions of L. Mumford, The City in History (Harcourt, 1961). Miscellaneous other titles on urbanism include G. Clay, Close-Up: How to Read the American City (Univ. of Chicago Press, 1980); A.H. Hawley, Urban Society: An Ecological Approach (Wiley, 2nd edn., 1980); F.S. Chapin, Jr., and E.J. Kaiser, Urban Land Use Planning (Univ. of Illinois Press, 3rd edn., 1979); N. Weinberg, Preservation in American Towns and Cities (Westview, 1979); R.P. Appelbaum, Size, Growth and U.S. Cities (Praeger, 1978); D.C. Perry and A.J. Watkins, eds., The Rise of the Sunbelt Cities (Beverly Hills CA: Sage Publications, Urban Affairs Annual Reviews 14, 1977); G. Sternlieb and J.W. Hughes, "New Regional and Metropolitan Realities of America," Journal of the American Institute of Planners 43 (1977); R.M. Christie, "The City in the American Industrial System," Growth and Change 7 (1976); B. Schwartz, ed., The Changing Face of the Suburbs (Univ. of Chicago Press, 1976); T.M. Stanback and R.V. Knight, Suburbanization and the City (Allanheld, Osmun, 1976); G. Sternlieb and J.W. Hughes, Post-Industrial America: Metropolitan Decline and Interregional Job Shifts (New Brunswick NJ: Center for Urban Policy Research, 1975); N.P. Gist and S.F. Fava,

Urban Society (Harper and Row, 6th edn., 1974); J.W. Hughes, ed., Suburbanization Dynamics and the Future of the City (New Brunswick NJ: Center for Urban Policy Research, 1974); A. Turner, "New Communities in the United States," Town Planning Review 45 (1974); J.R. Passonneau and R.S. Wurman, Urban Atlas: 20 American Cities (MIT Press, 1966), comprising selected social and economic data represented by statistical symbols in colour on detailed black-and-white street plans; K. Lynch, The Image of the City (MIT Press, 1960); and L. Wirth, "Urbanism as a Way of Life," American Journal of Sociology 44 (1938). The Journal of the American Planning Association (quarterly) carries many relevant articles and reviews, as well as massive bibliographical listings in urban studies and planning. In concluding this section on urban studies, the writer must stress the utility of many references in other parts of this chapter - particularly the general geographies cited on pp. 135-137, the references on social and economic geography on pp 143-155, and the citations grouped by regions in the discussion that follows.

## Regions, States, and Localities

Despite the present nation-wide uniformity of certain aspects of its culture and economy, the United States historically has possessed a strongly regional quality and in many ways does so today. A large share of the geographical and related literature on the country consists of studies of individual regions, sections, states, and localities. In recent years there has been an important emphasis on studies of states or regional groups of states by geographers, historians, and journalists. Examples include a series of geographies of individual states published by Westview Press, another series of state geographies published by Kendall-Hunt, a series of state histories published by Norton, and the several volumes on groups of states written by a journalist, Neal R. Peirce and published by Norton. Comprehensive reviews of the Norton histories and the Peirce volumes are available, respectively, in the American Historical Review 85 (June, 1980) and the AAG Annals 66 (September, 1976). It may be noted that the fine series of state guides published by the Federal Writers' Project in the American Guide Series (1937-1952) remains a huge mine of data on states and localities. Space precludes individual citation of volumes in the Norton histories and the American Guide Series, but the state geographies and the books by Peirce will be cited under the appropriate regions.

In the subsequent discussion, the references will be grouped under the headings of Northeast, South, Middle West, and West. For the purpose here, the Northeast is defined as the six New England States and the Middle Atlantic States (New York, Pennsylvania, New Jersey, Maryland, and Delaware), plus the District of Columbia. The South comprises Virginia,

West Virginia, Kentucky, Tennessee, Arkansas, Oklahoma, the Carolinas, Georgia, and the states that touch the Gulf of Mexico. The Middle West, here synonymous with the U.S. Census Bureau's North Central States, stretches from Ohio to Missouri, Kansas, Nebraska, and the Dakotas. The remainder of the United States lies in the West, which commences on the east with the states of New Mexico, Colorado, Wyoming, and Montana. The bibliographer regrets any violence done to customary regional conceptions by an indexing scheme using state lines as regional boundaries. In practice, a good share of all citations can be conveniently assigned to a particular state, and most of the rest can be accommodated within one of the four regional blocks of states. General citations for the Appalachians and the Interior Highlands (Ozarks and Ouachitas) will be arbitrarily assigned to the South in most cases, and corresponding citations for the Great Plains will be placed under the West.

## The Northeast

Considering the enormous importance of the Northeast to the United States throughout the country's history, as well as the intricacy of the region's geography, history, culture, and social relations, the number of geographical studies that have a Northeastern setting is curiously small. Nor is there an overwhelming volume of truly relevant publications in cognate fields. Far more has been written about the South or the West. The relative lack of geographical studies on the Northeast is doubtless related to the long neglect of geography as an academic discipline by many of the leading universities and colleges in the region. By contrast, the discipline has been well developed in many leading institutions of the Middle West, West and South as well as in a few North-eastern institutions. Readers wanting to pursue this facet of American educational and intellectual history will find material of interest in J.C. Hudson, ed., "Seventy-Five Years of American Geography," Special Issue, AAG *Annals* 69 (March, 1979) and in P.E. James and G.J. Martin, *The Association of American Geographers: The First Seventy-Five Years, 1904-1979* (AAG, 1978).

Undoubtedly the best-known work on the Northeast by a geographer is J. Gottman, *Megalopolis: The Urbanized Northeastern Seaboard of the United States* (Twentieth Century Fund, 1961), some aspects of which are summarized in Gottman, "Megalopolis or the Urbanization of the Northeastern Seaboard," *Economic Geography* 47 (1957). For an extended review of Gottman's book, see H.J. Nelson, "Megalopolis and New York Metropolitan Region: New Studies of the Urbanized Eastern Seaboard," AAG Annals 52 (1962). See also L.A. Swatridge, *The Bosnywash Megalopolis* (McGraw-Hill Ryerson, 1971). Miscellaneous geographical references on the Northeast as a whole include L.M. Alexander, *The Northeastern United States*

(Van Nostrand Reinhold, 2nd edn., 1976); G.M. Lewis, "Levels of Living in the North-eastern United States c. 1960: A New Approach to Regional Geography," IBG Transactions 45 (1968); W. Zelinsky, "Generic Terms in the Place Names of the North-eastern United States," AAG Annals 45 (1955); L.E. Klimm, "The Empty Areas of the Northeastern United States," Geographical Review 44 (1954); and G.T. Trewartha, "Types of Rural Settlement in Colonial America," Geographical Review 36 (1946). An important cognate reference is G. Sternlieb and J.W. Hughes, eds., Revitalizing the Northeast: Prelude to an Agenda (New Brunswick NJ: Center for Urban Policy Research, 1978). Four Northeastern states- New York, Pennsylvania, New Jersey, and Massachusetts - are among the "Ten Great States" discussed in N.R. Peirce, The Megastates of America (Norton, 1972).

Selected titles by geographers on New England include B. Wallach, "The Potato Landscape: Aroostock County, Maine," Landscape 23 (1979); M.P. Conzen and G.K. Lewis, "Boston: A Geographical Portrait," in J.S. Adams, ed., Contemporary Metropolitan America, op cit., Vol.1 (1976); D.R. McManis, Colonial New England: A Historical Geography (OUP, 1975), and, McManis, European Impressions of the New England Coast, 1497-1620 (Univ. of Chicago, Geography Research Paper 139, 1972); G.K. Lewis, "Population Change in Northern New England," in J.F. Hart, ed., Regions of the United States, op.cit. (1972); D. Ward, "The Emergence of Central Immigrant Ghettoes in American Cities: 1840-1920",AAG Annals 58 (1968); R.C. Estall, New England: A Study in Industrial Adjustment (Praeger, 1966); A.N. Strahler, A Geologist's View of Cape Cod (Natural History Press, 1966); W.H. Wallace, "Merrimack Valley Manufacturing: Past and Present," Economic Geography 37 (1961); E.C. Higbee, "The Three Earths of New England," Geographical Review 42 (1952); J.H. Thompson and E.C. Higbee, New England Excursion Guidebook (Washington, D.C.: 17th International Geographical Congress, 1952); E. Scofield, "The Origin of Settlement Patterns in Rural New England," Geographical Review 28 (1938); J.K. Wright, ed., New England's Prospect (American Geographical Society, 1933); and P.E. James, "The Blackstone Valley....," AAG Annals 19 (1929). A few cognate references of special interest are H.S. Russell, Indian New England before the Mayflower (Univ. Press of New England, 1980); N.R. Peirce, The New England States: People, Politics and Power in the Six New England States (Norton, 1976); H.S. Russell, A Long, Deep Furrow: Three Centuries of Farming in New England (Univ. Press of New England, 1976); R. Bearse, ed., Massachusetts: A Guide to the Pilgrim State (Houghton Mifflin, 2nd edn. rev., 1971); W.M. Whitehill, Boston: A Topographical History (Harvard Univ. Press, 1968); S.E. Morison, The Maritime History of Massachusetts, 1783-1860 (Houghton Mifflin, 1941); and L.M. Rosenberry, The Expansion of New England: The Spread of New England Settlement and Institutions to the Mississippi River, 1620-1865 (Houghton Mifflin, 1909).

Outstanding among the publications by geographers on the Middle Atlantic States is J.H. Thompson, ed., Geography of New York State (Syracuse Univ. Press, 2nd edn., 1977). Other titles include J.E. Dilisio, Maryland: A Geography (Westview, forthcoming); C. Stansfield, New Jersey: A Geography (Westview, forthcoming); R.A. Cybriwsky, ed., The Philadelphia Region: Selected Essays and Field Trip Itineraries (AAG, 1979); W.K. Crowly, "Old Order Amish Settlement: Diffusion and Growth," AAG Annals 68 (1978); W. Zelinsky,"The Pennsylvania Town: An Overdue Geographical Account," Geographical Review 68 (1978); F.S. Kelland and M.C. Kelland, New Jersey : Garden or Suburb?: A Geography of New Jersey (Kendall/Hunt, 1978); individual city studies in J.S. Adams, ed., Contemporary Metropolitan America, op. cit. (1976), as follows: Vol.1, G.W. Carey, "A Vignette of the New York-New Jersey Metropolitan Region," D.R. Meyer, "From Farm to Factory to Urban Pastoralism: Urban Change in Central Connecticut, and P.O. Muller, K.C. Meyer, and R.A. Cybriwsky, "Metropolitan Philadelphia: A Study of Conflicts and Cleavages"; Vol. 2, S. Olson, "Baltimore"; Vol. 3, P.H. Vernon and O. Schmidt, "Metropolitan Pittsburgh: Old Trends and New Directions"; and Vol. 4, J-C. M. Thomas, "Washington, D.C."; J.E. Brush and G.W. Carey, eds., Guidebook for Field Excursions: New York City Meetings (AAG, 1976); E.F. Bergman and T.W. Pohl, A Geography of the New York Metropolitan Region (Kendall/Hunt, 1975); C.V. Earle, The Evolution of a Tidewater Settlement System: All Hallow's Parish, Maryland, 1650-1783 (Univ. of Chicago, Geography Research Paper 170, 1973); P.O. Wacker, Land and People: A Cultural Geography of Preindustrial New Jersey: Origins and Settlement Patterns (Rutgers Univ. Press, 1975); D. Ley, The Black Inner City as Frontier Outpost: Images and Behavior of a Philadelphia Neighborhood (AAG Monograph 7, 1974); J.T. Lemon, The Best Poor Man's Country: A Geographical Study of Early Southeastern Pennsylvania (Johns Hopkins Univ. Press, 1972); P.F. Lewis, "Small Town in Pennsylvania," in J.F. Hart, ed., Regions of the United States, op. cit. (1972); M.R. Greenberg et al., "A Geographical Systems Analysis of the Water Supply Networks of the New York Metropolitan Region," Geographical Review 61 (1971); R.A. Harper and F.O. Ahnert, Introduction to Metropolitan Washington (AAG, 1968); AAG, Nine Geographical Field Trips in the Washington, D.C. Area (1968); L. Durand, Jr., "The Historical and Economic Geography of Dairying in the North Country of New York State," Geographical Review 57 (1967); P. Hall, "New York," in Hall, The World Cities (Weidenfeld and Nicolson, 2nd edn., 1977); and G.F. Deasy and P.R. Griess, "Effects of a Declining Mining Economy on the Pennsylvania Anthracite Region," AAG Annals 55 (1965). A brief selection of cognate references includes M.D. Lopez, New York: A Guide to Information and Reference Sources (Scarecrow Press, 1980); R.T.T. Forman, ed., Pine Barrons: Ecosystem and Landscape (Academic, 1979); and M. Hall, ed., New York Metropolitan Region Study, 10 vols. (Harvard Univ. Press, 1959-1960).

The South

Geographical literature on the American South is not voluminous when compared to the flood of writing on this region by historians, novelists, and journalists. But it is certainly more extensive than comparable literature on the Northeast, and it includes some very perceptive and scholarly work. We begin with a survey of regional, historical, and topical writing by geographers on the South as a whole, or on regions touching more than one Southern state. This is followed by a discussion of geographical writing on individual states and localities. Finally, some especially meaningful citations are extracted from the immense literature on the South in cognate fields.

A great many references in earlier parts of this chapter contain valuable information on the South. In this paragraph we focus on general and miscellaneous titles by geographers not previously cited. Agricultural geography is surveyed separately, as are industrial and urban geography, and cognate titles of all types. Selected titles conveying broad general perspectives on the region's geography include K.B. Raitz and R. Ulack, "Cognitive Maps of Appalachia," Geographical Review 71 (1981); A.B. Cruickshank, "Development of the Deep South: A Reappraisal," Scottish Geographical Magazine 96 (1980); M.S. Meade, "The Rise and Demise of Malaria: Some Reflections on Southern Settlement and Landscape," Southeastern Geographer 20 (1980); C.S. Aiken, "Faulkner's Yoknapatawpha County: A Place in the American South," Geographical Review 69 (1979), and, Aiken, "Faulkner's Yoknapatawpha County: Geographical Fact into Fiction," Geographical Review 67 (1977); M.C. Prunty, "Two American Souths: The Past and the Future,"Southeastern Geographer 17 (1977); J.F. Hart, The South(Van Nostrand Reinhold, 2nd edn., 1976); H.D. Clout, "Les Appalaches: Une région américaine en crise," Annales de Géographie 458 (1974); G. Wilhelm, Jr., "The Mullein: Plant Piscicide of the Mountain Folk Culture," Geographical Review 64 (1974); J. Heyl, ed., The South: A Vade Mecum (AAG, 1973); T.A. Hartshorn, "The Spatial Structure of Socioeconomic Development in the Southeast, 1950-1960," Geographical Review 61 (1971); E.J.W. Miller, "The Ozark Culture Region as Revealed by Traditional Materials," AAG Annals 58 (1968); E.T. Price, "Root Digging in the Appalachians: The Geography of Botanical Drugs," Geographical Review 50 (1960); D.J. de Laubenfels, "Where Sherman Passed By," Geographical Review 47 (1957); R.C. West, "The Term 'Bayou' in the United States: A Study in the Geography of Place Names," AAG Annals 44 (1954); A.E. Parkins, The South: Its

Economic-Geographic Development (Wiley, 1938), and Parkins, "The Antebellum South: A Geographer's Interpretation," AAG Annals 21 (1931).

Titles by geographers on land use, food supply, agriculture, fisheries, and forestry include J.J. Winberry, "The Sorghum Syrup Industry, 1854-1975," Agricultural History 54 (1980); I.R. Manners, "The Persistent Problem of the Boll Weevil: Pest Control in Principle and in Practice," Geographical Review 69 (1979); C.V. Earle, "A Staple Interpretation of Slavery and Free Labor," Geographical Review 68 (1978); J.F. Hart, "Cropland Concentrations in the South," AAG Annals 68 (1978), also, Hart, "The Demise of King Cotton," AAG Annals 67 (1977), and "Land Rotation in Appalachia," Geographical Review 67 (1977); C.S. Aiken, "The Evolution of Cotton Ginning in the Southeastern United States," Geographical Review 63 (1973); J.R. Anderson, A Geography of Agriculture in the United States' Southeast (Geography of World Agriculture, No.2; Budapest: Akadémiai Kiadó, Publishing House of the Hungarian Academy of Sciences, 1973); J.S. Fisher, "Negro Farm Ownership in the South," AAG Annals 63 (1973); S.B. Hilliard, Hog Meat and Hoecake: Food Supply in the Old South, 1840-1860 (Southern Illinois Univ. Press, 1972); M.C. Prunty and C.S. Aiken, "The Demise of the Piedmont Cotton Region," in J.F. Hart, ed., Regions of the United States, op. cit. (1972); R.B. Lamb, The Mule in Southern Agriculture (Univ. of California Publications in Geography 15, 1963); H.R. Padgett, "The Sea Fisheries of the Southern United States," Geographical Review 53 (1963); M.C. Prunty, "The Woodland Plantation as a Contemporary Occupance Type in the South," Geographical Review 53 (1963), also, Prunty, "The Renaissance of the Southern Plantation," Geographical Review 45 (1955), and "Land Occupance in the Southeast: Landmarks and Forecast," Geographical Review 42 (1952).

Among the titles by geographers on industrial and urban geography are T. Klimasewski, "Corporate Dominance of Manufacturing in Appalachia," Geographical Review 68 (1978); C. Earle and R. Hoffman (a historian), "The Urban South: The First Two Centuries," in B.A. Brownell and D.R. Goldfield, eds., The City in Southern History (Kennikat, 1977); J. Kellogg, "Negro Urban Clusters in the Postbellum South," Geographical Review 67 (1977); J.R. McGregor, "A Delimitation of Manufacturing Regions within the Southeastern United States," in Department of Geography and Geology, Indiana State Univ., Professional Paper 9 (1977); R.E. Londsdale and C.E. Browning, "Rural-Urban Locational Preferences of Southern Manufacturers," AAG Annals 61 (1971); and J.F. Hart, "Functions and Occupational Structures of Cities of the American South," AAG Annals 45 (1955).

A considerable literature has accumulated on the geography of individual Southern states and localities. The selections that follow are arranged alphabetically by state.

They include N.G. Lineback, ed., Atlas of Alabama (Univ. of Alabama Press, 1973); J.A. Tower, "Cotton Change in Alabama, 1879-1946," Economic Geography 26 (1950);E.J.W. Miller, "The Naming of the Land in the Arkansas Ozarks: A Study in Culture Processes," AAG Annals 59 (1969); D.B. Longbrake and W.W. Nichols, Jr., "Sunshine and Shadows in Metropolitan Miami," Florida , in J.S. Adams, ed., Contemporary Metropolitan America, op. cit., Vol. 4 (1976); W.T. Mealor, Jr., and M.C. Prunty, "Open-Range Ranching in Southern Florida," AAG Annals 66 (1976); R.B. Marcus and E.A. Fernald, Florida: A Geographical Approach (Kendall/Hunt, 1974); R. Wood and E.A. Fernald The New Florida Atlas: Patterns of the Sunshine State (Tampa FL: Trend Publications, 1974); L.A. Eyre, "Land Reclamation and Settlement of the Florida Everglades," Geography 57 (1972); E. Raisz et al., Atlas of Florida (Univ. of Florida Press, 1964); J.F. Hart, "Land Use Change in a Piedmont County," AAG Annals 70 (1980); T.A. Hartshorn et al., "Metropolis in Georgia: Atlanta's Rise as a Major Transaction Center," in J.S. Adams, ed., Contemporary Metropolitan America, op. cit., Vol. 4 (1976); D.E. Bierman, ed., An Introduction to the Louisville Region: Selected Essays (AAG, 1980); R.R. Dilamarter and W.L. Hoffman, eds., AAG Louisville Meeting: Field Trip Guide, 1980 (AAG, 1980); K.B. Raitz, "The Barns of Barren County," Landscape 22 (1978); P.P. Karan and C. Mather, eds., Atlas of Kentucky (Univ. Press of Kentucky, 1977); P.C. Smith and K.B. Raitz, "Negro Hamlets and Agricultural Estates in Kentucky's Inner Bluegrass," Geographical Review 64 (April 1974); C.O. Sauer, Geography of the Pennyroyal (1927), one of a series of volumes on the geography of Kentucky published by the Kentucky Geological Survey, 1923-1928; E.C. Semple, "The Anglo-Saxons of the Kentucky Mountains," Geographical Journal 17 (1901); S.B. Hilliard, ed., Man and Environment in the Lower Mississippi Valley. (Louisiana State Univ., Geoscience and Man, Vol. 19, 1978); R.H. Kesel and R.A. Sauder, A Field Guidebook for Louisiana (AAG 1978) P.F. Lewis, "New Orleans: The Making of an Urban Landscape," in J.S. Adams, ed., Contemporary Metropolitan America , Vol. 2 (Ballinger, 1976); M.L. Comeaux Atchafalaya Swamp Life: Settlement and Folk Occupations (Louisiana State Univ., Geoscience and Man, Vol. 2, 1972); F.B. Kniffen, Louisiana: Its Land and People (Louisiana State Univ. Press, 1968); The Mississippi Geographer (annual); J.O. McKee, Mississippi: A Geography (Westview, forthcoming); R.D. Cross et al., Atlas of Mississippi (Univ. of Mississippi Press, 1974); M. Lowry II, "Population and Race in Mississippi, 1940-1960," AAG Annals 61 (1971), and Lowry, "Race and Socioeconomic Well-Being: A Geographical Analysis of the Mississippi Case," Geographical Review 60 (1970); O. Gade and H.D. Stillwell, North Carolina: A Geography (Westview, forthcoming); J.F Hart and E.L. Chestang, "Rural Revolution in East Carolina," Geographical Review 68 (1978); J.W. Clay et al., eds., North Carolina Atlas: Portrait of a Changing Southern State (Univ. of North Carolina Press, 1976); H.R. Merrens, Colonial North Carolina in the Eighteenth Century: A Study in Historical Geography (Univ. of North Carolina Press, 1964); J.W. Morris et al., Historical Atlas of Oklahoma (Univ. of Oklahoma Press, 2nd edn., 1976); C.F. Kovacik and J.J. Winberry, South Carolina: A Geography (Westview, forthcoming); R.O. Fullerton and J.B. Ray, eds., Tennessee: Geographical Patterns and Regions (Kendall/Hunt, 1977); T.G. Jordan, Texas: A Geography (Westview, forthcoming), also, Jordan, "Perceptual Regions in Texas," Geographical Review 68 (1978), Texas Log Buildings: A Folk Architecture (Univ. of Texas Press, 1978), and "Early Northeast Texas and the Evolution of Western Ranching," AAG Annals 67 (1977); S.A. Arbingast et al., Atlas of Texas (Univ. of Texas at Austin, Bureau of Business Research, rev. edn., 1976); D. Conway et al., "The Dallas-Fort Worth Region," in J.S. Adams, ed., Contemporary Metropolitan America, Vol. 4, (Ballinger, 1976), and M.E. Palmer and M.N. Rush, "Houston," in the same volume of Adam's important work; K. Weigand, Stadtgeographische Studien in Südwesttexas (Wiesbaden: Steiner, 1973); T.G. Jordan "Population Origin Groups in Rural Texas," Map Supplement 13, AAG Annals 60 (1970), and Jordan, "The Texas Appalachia," AAG Annals 60 (1970); D.W. Meinig, Imperial Texas: An Interpretive Essay in Cultural Geography (Univ. of Texas Press, 1969); T.G. Jordan, "The Imprint of the Upper and Lower South on Mid-Nineteenth-Century Texas," AAG Annals 57 (1967), and Jordan, German Seed in Texas Soil: Immigrant Farmers in Nineteenth-Century Texas (Univ. of Texas Press, 1966); J. Fonseca and G.H. Stopp, Virginia: A Geography (Westview, forthcoming); C.V. Earle, "Environment, Disease, and Mortality in Early Virginia," in T.W. Tate and D.L. Ammerman, eds., The Chesapeake in the Seventeenth Century: Essays on Anglo-American Society (Univ. of North Carolina Press, 1979); R.D. Mitchell, Commercialism and Frontier: Perspectives on the Early Shenandoah Valley (Univ. Press of Virginia, 1977), and Mitchell, "The Shenandoah Valley Frontier," AAG Annals 62 (1972); J. Gottmann, Virginia in Our Century (Univ. Press of Virginia, 1969); T.D. Hankins, West Virginia: A Geography (Westview, forthcoming); H.G. Adkins et al., West Virginia and Appalachia: Selected Readings (Kendall/Hunt, 1977); R.C. Estall, "Appalachian State: West Virginia as a Case Study in the Appalachian Regional Development Problem," Geography 53 (1968); and J.H. Wheeler, Jr., Land Use in Greenbrier County. West Virginia (Univ. of Chicago, Geography Research Paper 15, 1950).

In the foregoing discussion, the titles of many valuable articles in The Southeastern Geographer (semiannual) have been omitted due to limited space. Readers interested in the South are urged to consult the files of this fine regional journal.

A selection of titles in cognate fields includes G.M. Fredrickson, White Supremacy: A Comparative Study in American and South African History (OUP, 1981); W.W. Philliber et al., The Invisible Minority: Urban Appalachians (Univ. Press of Kentucky, 1981); N.M. Blake, Land into Water-Water into Land: A History of Water Management in Florida (Univ. Presses of Florida, 1980); F. McDonald and G. McWhiney, "The South from Self-Sufficiency to Peonage: An Interpretation," American Historical Review 85 (1980), and E. Pessen, "How Different from Each Other were the Antebellum North and South?", ibid.; R.P. Steed et al., Party Politics in the Deep South (Praeger, 1980); G.L. Robson, Jr., and R.V. Scott, eds., "Southern Agriculture since the Civil War: A Symposium," special issue, Agricultural History 53 (January, 1979); D.C. Roller and R.W. Twyman, eds., Encyclopedia of Southern History (Louisiana State Univ. Press, 1979); J.C. Bonner, "House and Landscape Design in the Antebellum South," Landscape 21 (1977); J. Bass and W. DeVries, The Transformation of Southern Politics: Social Change and Political Consequence Since 1945 (Basic Books, 1976); L.F. Wheat, Urban Growth in the Nonmetropolitan South (Lexington Books, 1976); C. Eaton, A History of the Old South (Macmillan, 3rd edn., 1975); H. Glassie, Folk Housing in Middle Virginia: A Structural Analysis of Historic Artifacts (Univ. of Tennessee Press, 1975); N.R. Peirce, The Border South States (Norton, 1975); K. Sale, Power Shift: The Rise of the Southern Rim and its Challenge to the Eastern Establishment (Random House, 1975); L.J. Carter, The Florida Experience: Land and Water Policy in a Growth State (Johns Hopkins Univ. Press for Resources for the Future, 1974); J. Egerton, The Americanization of Dixie: The Southernization of America (Harper and Row, 1974); N.R. Peirce, The Deep South States of America (Norton, 1974); D.B. Dodd and W.S. Dodd, Historical Statistics of the South 1790-1970 (Univ. of Alabama Press, 1973); W.H. Stephenson and E.M. Coulter, eds., A History of the South, 10 vols. (Louisiana State Univ. Press, 1947-1973); N.M. Hansen, Location Preference, Migration, and Regional Growth: A Study of South and Southwest United States (Praeger, 1973); H.B. Ayers and T.H. Naylor, eds., You Can't Eat Magnolias (McGraw-Hill, 1972); R. Coles, Farewell to the South (Atlantic/Little, Brown, 1972); N.R. Peirce, The Megastates of America (Norton, 1972); F.B. Simkins and C.P. Roland, A History of the South (Random House, 4th edn., 1972); D.S. Walls and J.B. Stephenson, eds., Appalachia in the Sixties (Univ. Press of Kentucky, 1972); E. Wigginton, ed., The Foxfire Book: Appalachian folk culture (Anchor Press/ Doubleday, 1972), and by the same author and publisher, Foxfire 2 (1973), Foxfire 3 (1975), Foxfire 4 (1977), and Foxfire 5 (1979); M.L. Billington, The American South: A Brief History (Scribner's, 1971); W.N. Parker, ed., "The Structure of the Cotton Economy of the Antebellum South," special issue, Agricultural History 44 (January, 1970);

E. Caldwell, Deep South (Weybright and Talley, 1968); C.V. Woodward, The Burden of Southern History (Louisiana State Univ. Press, rev. edn., 1968); T.D. Clark and A.D. Kirwan, The South Since Appomattox: A Century of Regional Change (OUP, 1967); F.E. Vandiver, ed., The Idea of the South (Univ. of Chicago Press, 1964); H.M. Caudill, Night Comes to the Cumberlands: A Biography of a Depressed Area (Atlantic/Little, Brown, 1963), together with sequels by Caudill entitled My Land Is Dying (Dutton, 1971) and The Watches of the Night (Atlantic/Little,Brown, 1976); W.P. Cumming, The Southeast in Early Maps (Univ. of North Carolina Press, 2nd edn., 1962); T.R. Ford, ed., The Southern Appalachian Region: A Survey (Univ. of Kentucky Press, 1962); F.B. Simkins, "The South",in M. Jensen, ed., Regionalism in America (Univ. of Wisconsin Press, 1951); V.O. Key, Jr., Southern Politics in State and Nation (Knopf, 1949); G.E. McLaughlin and S. Robock, Why Industry Moves South (National Planning Association, 1949); J. Agee and W. Evans, Let Us Now Praise Famous Men: Three Tenant Families (Houghton Mifflin, 1941); W.J. Cash, The Mind of the South (Knopf, 1941); H.W. Odum, Southern Regions of the United States (Univ. of North Carolina Press, 1936); W.T. Couch, ed., Culture in the South (Univ. of North Carolina Press, 1934); and R.P. Vance, Human Geography of the South: A Study in Regional Resources and Human Adequacy (Univ. of North Carolina Press, 1932).

The Middle West

There is a comparative dearth of recent articles and books on the geography of the Middle West. This situation seems anomalous in view of the high reputation of the region for advanced work in academic geography. Whatever the causes may be, the searcher is advised not to expect a large amount of writing by geographers bearing publication dates in the quarter-century from 1955 to 1980. This is not to impugn the value of fine contributions by J.F. Hart, H.B. Johnson, H.M. Mayer, and others cited below. We first survey titles by geographers and then list a few references in cognate fields.

Selected works by geographers include L.M. Sommers, Michigan: A Geography (Westview, forthcoming); M.D. Rafferty, Missouri: A Geography (Westview, forthcoming); W.A. Schroeder, Presettlement Prairies of Missouri (Jefferson City MO: Missouri Department of Conservation, Natural History Series No.2, 1981, with map by the same title published separately at 1:500,000); M.D. Rafferty, The Ozarks: Life and Land (Univ. of Oklahoma Press, 1980); L. Dillon and E.E. Lyon, Indiana: Crossroads of America (Kendall/Hunt, 1978); H.B. Johnson, "Perceptions and Illustrations of the American Landscape in the Ohio Valley and the Midwest," in This Land of Ours: The Acquisition and Disposition of the Public Domain (Indiana Historical Society, 1978); R.E. Nelson, ed., Illinois: Land and Life in the Prairie State (Kendall/Hunt, 1978); H. Self,

Environment and Man in Kansas: A Geographical Analysis (Lawrence KS: Regents Press of Kansas, 1978); A.M. Davis, "The Prairie-Deciduous Forest Ecotone in the Upper Middle West," AAG Annals 67 (1977); J.A. Jakle, Images of the Ohio Valley: A Historical Geography of Travel, 1740-1860 (OUP, 1977); M.P. Lawson, K.F. Dewey, and R.E. Neild, Climatic Atlas of Nebraska (Univ. of Nebraska Press, 1977); R.E. Lonsdale, ed., Economic Atlas of Nebraska (Univ. of Nebraska Press, 1977); R.A. Santer, Michigan: Heart of the Great Lakes (Kendall/Hunt, 1977); W.A. Schroeder, Bibliography of Missouri Geography, (Univ. of Missouri-Columbia Extension Division, 1977); J.H. Williams and D. Murfield, eds., Agricultural Atlas of Nebraska (Univ. of Nebraska Press, 1977); J.S. Adams, ed., Contemporary Metropolitan America, op. cit., Vol. 3 (1976), chapters as follows: R. Abler, J.S. Adams, and J.R. Borchert, "The Twin Cities of St. Paul and Minneapolis," B.J.L. Berry et al., "Chicago: Transformations of an Urban System," D.K. Ehrhardt, "The St. Louis Daily Urban System," H.M. Mayer and T. Corsi, "The Northeastern Ohio Urban Complex," R. Sinclair and B. Thompson, "Detroit"; I. Cutler, Chicago: Metropolis of the Mid-Continent (Kendall/Hunt, 2nd edn., 1976); R.L. Gerlach, Immigrants in the Ozarks: A Study in Ethnic Geography (Univ. of Missouri Press, 1976); J.C. Hudson, "Migration to an American Frontier" [North Dakota] , AAG Annals 66 (1976); H.B. Johnson, Order Upon the Land: The U.S. Rectangular Land Survey and the Upper Mississippi Country (OUP, 1976); D.K. Meyer, "Illinois Culture Regions at Mid-Nineteenth Century," Bulletin of the Illinois Geographical Society 18 (1976);E.K. Muller, "Selective Urban Growth in the Middle Ohio Valley, 1800-1860," Geographical Review 66 (1976); C. Mather, J.F. Hart, H.B. Johnson, and R. Matros, Upper Coulee Country (Prescott WI: Trimbelle Press, 1975); M.D. Sublett, Farmers on the Road: Interfarm Migration and the Farming of Noncontiguous Lands in Three Midwestern Townships, 1939-1969 (Univ. of Chicago, Geography Research Paper 168, 1975); C.B. McIntosh, "Forest Lieu Selections in the Sand Hills of Nebraska," AAG Annals 64 (1974); A.H. Robinson and J.B. Culver, Atlas of Wisconsin (Univ. of Wisconsin Press, 1974); J.C. Hudson, "Two Dakota Homestead Frontiers," AAG Annals 63 (1973); J.F. Hart, "The Middle West," in Hart, ed., Regions of the United States, op. cit. (1972), a fine essay with which to begin a study of the region; W.E. Kiefer, "An Agricultural Settlement Complex in Indiana," AAG Annals 62 (1972); K.B. Raitz and C. Mather, "Norwegians and Tobacco in Western Wisconsin," AAG Annals 61 (1971); J.S. Adams, "Residential Structure of Midwestern Cities," AAG Annals 60 (1970); R. Sinclair, The Face of Detroit: A Spatial Synthesis (Wayne State University Press, 1970); H.M.Mayer and R.C. Wade, Chicago: Growth of a Metropolis (Univ. of Chicago Press, 1969); W.E. Akin, The North Central United States (Van Nostrand, 1968); J.F. Hart, "Field Patterns in Indiana," Geographical Review 58 (1968); National Council for Geographic

Education, "Geography Through Maps" (1967), Special Publications as follows: D.E. Christensen and R.A. Harper, eds., The Mississippi-Ohio Confluence Area, J.W. Conoyer, The Saint Louis Gateway, H.B. Kircher, The Southern Illinois Prairies, and W.A. Schroeder, The Eastern Ozarks; M Kaups, "Finnish Place Names in Minnesota: A Study in Cultural Transfer," Geographical Review 56 (1966); W.M. Kollmorgen and D.S. Simonett, "Grazing Operations in the Flint Hills-Bluestem Pastures of Chase County, Kansas," AAG Annals 55 (1965); T.G. Jordan, "Between the Forest and the Prairie," Agricultural History 38 (1964); G.R. Lewthwaite, "Wisconsin and the Waikato: A Comparison of Dairy Farming in the United States and New Zealand," AAG Annals 54 (1964); D.R. McManis, The Initial Evaluation and Settlement of the Illinois Prairies, 1815-1840 (Univ. of Chicago, Geography Research Paper 94 1964); H.M. Mayer, "Politics and Land Use: The Indiana Shoreline of Lake Michigan," AAG Annals 54 (1964); A.K. Philbrick, "The Nodal Water Region of North America," Canadian Geographer 8 (1964); J.E. Spencer and R.J. Horvath, "How Does an Agricultural Region Originate? [U.S. Corn Belt] AAG Annals 53 (1963); C.O. Sauer, "Homestead and Community on the Middle Border," Landscape 12 (1962), reprinted in Landscape 20 (1976) and in J. Leighly, ed., Land and Life: A Selection from the Writings of Carl Ortwin Sauer (Univ. of California Press, 1967); J.W. Brownell, "The Cultural Midwest," Journal of Geography 59 (1960); H.B. Johnson, "Rational and Ecological Aspects of the Quarter Section: An Example from Minnesota," Geographical Review 47 (1957); H.M. Mayer, The Port of Chicago and the St. Lawrence Seaway (Univ. of Chicago, Geography Research Paper 49, 1957); J.C. Weaver, L.P. Hoag, and B.L. Fenton, "Livestock Units and Combination Regions in the Middle West," Economic Geography 32 (1956); J.E. Collier, "Geographic Regions of Missouri," AAG Annals 45 (1955); J.H. Garland, ed., The North American Midwest: A Regional Geography (Wiley, 1955); H.M. Mayer, "Prospects and Problems of the Port of Chicago," Economic Geography 31 (1955); J.C. Weaver, "Changing Patterns of Cropland Use in the Middle West," Economic Geography 30 (1954), and Weaver, "Crop-Combination Regions in the Middle West," Geographical Review 44 (1954); J.E. Collier, Geography of the Northern Ozark Border Region in Missouri (University of Missouri Studies 26, 1953); R.S. Thoman, The Changing Occupance Pattern of the Tri-State Area: Missouri, Kansas, and Oklahoma (Univ. of Chicago, Geography Research Paper 31, 1953); H.B. Johnson, "The Location of German Immigrants in the Middle West," AAG Annals 41 (1951); H.A. Gleason, "The Vegetational History of the Middle West," AAG Annals 12 (1922); and C.O. Sauer, The Geography of the Ozark Highland of Missouri (Geographic Society of Chicago Bulletin 7; Univ. of Chicago Press, 1920). Space limitations preclude the listing of a great many titles prior to 1950, including a number of early studies in the Geography Research

Series of the University of Chicago. The searcher is advised to consult standard bibliographical aids discussed early in this chapter, as well as bibliographies and reference footnotes in titles cited above.

A few cognate references on the Middle West include N.R. Peirce and J. Keefe, The Great Lakes States of America: People, Politics, and Power in the Five Great Lakes States (Norton, 1980); N.R. Peirce, The Great Plains States of America (Norton, 1973); A.G. Bogue, From Prairie to Corn Belt: Farming on the Illinois and Iowa Prairies in the Nineteenth Century (Univ. of Chicago Press, 1963); E.W. McMullen, "The Term 'Prairie' in the United States," Names 5 (1957); J.C. Malin, The Grassland of North America: Prolegomena to its History (Lawrence KS: The Author, 1956), see also, R.G. Bell, "James C. Malin and the Grasslands of North America," Agricultural History 46 (1972); C.C. Taylor, "The Corn Belt," in Taylor et al., Rural Life in the United States (Knopf, 1955); L.E. Atherton, Main Street on the Middle Border (Univ. of Indiana Press, 1954); and T.J. Wertenbaker, "The Molding of the Middle West," American Historical Review 53 (1948).

The West (including general references on the Great Plains).

Writings on the American West by geographers are very numerous, and there is a superabundant literature in cognate fields. However, this body of material tends to cluster markedly around certain places and topics such as California; water resources; ranching; mining; exploration and the fur trade in the nineteenth century; and Mormonism. Titles by geographers will be grouped regionally under (a) the West in general, (b) the Great Plains in general, (c) the states and physical areas of the interior West (Arizona, Colorado, Idaho, Montana, Nevada, New Mexico, Utah, and Wyoming), and (d) the Pacific states (Alaska, California, Hawaii, Oregon, and Washington).

No full-scale geography of the entire West has yet appeared. Readers desiring a summary view can piece it together from appropriate sections of the general geographies cited on pp. 131-137. Also very helpful is J.F. Hart, ed., Regions of the United States, op. cit., which offers several substantial and well-documented articles on the West. The individual articles are cited in appropriate places below. Selected titles on the West in general, commencing with those of most recent date, include T.G. Jordan, Trails to Texas: Southern Roots of Western Cattle Ranching (Univ. of Nebraska Press, 1981); J.R. Gibson, "Russian Expansion in Siberia and America," Geographical Review 70 (1980); D.J. Wishart, The Fur Trade of the American West, 1807-1840: A Geographical Synthesis (Univ. of Nebraska Press, 1979); J. Leighly, "Town Names of Colonial New England in the West," AAG Annals 68 (1978); T.R. Vale, "The Sagebrush Landscape," Landscape 22 (1978); J.R. Gibson, Imperial Russia in Frontier America:

The Changing Geography of Supply of Russian America, 1784-1867 (OUP, 1976); J.L. Allen, Passage Through the Garden: Lewis and Clark and the Image of the American Northwest (Univ. of Illinois Press, 1975), and Allen, "An Analysis of the Exploratory Process: The Lewis and Clark Expedition of 1804-1806," Geographical Review 62 (1972); the highly valuable interpretation in D.W. Meinig, "American Wests: Preface to a Geographical Introduction," in J.F. Hart, ed., Regions of the United States, op. cit. (1972); C.W. Booth, The Northwestern United States (Van Nostrand Reinhold, 1971); J.W. Morris, The Southwestern United States (Van Nostrand Reinhold, 1970); J. Humlum, Water Development and Water Planning in the Southwestern United States (Humanities Press, 1969); F. Quinn, "Water Transfers: Must the American West Be Won Again?", Geographical Review 58 (1968); J.E. Vance, Jr., "The Oregon Trail and Union Pacific Railroad: A Contrast in Purpose," AAG Annals 51 (1961); M.W. Mikesell, "Comparative Studies in Frontier History," AAG Annals 51 (1961); J. Leighly, "John Muir's Image of the West," AAG Annals 48 (1958); C.M. Zierer, ed., California and the Southwest (Wiley, 1956); and E.W. Gilbert, The Exploration of Western America, 1800-1850 (Cambridge Univ. Press, 1933).

General references by geographers on the Great Plains include J.C. Hudson, "Great Plains, U.S.A.: The 1980s," in J. Rogge, ed., The Prairies and Plains: Prospects for the 80s, Manitoba Geographical Studies 7 (1981); B.W. Blouet and F.C. Luebke, eds., The Great Plains: Environment and Culture (Univ. of Nebraska Press, 1979); R.K. Sutton, "Circles on the Plains: Center Pivot Irrigation," Landscape 22 (1977); M.J. Bowden, "The Great American Desert in the American Mind: The Historiography of a Geographical Notion," in D. Lowenthal and M.J. Bowden, eds., Geographies of the Mind (OUP, 1976); B.W. Blouet and M.P. Lawson, eds., Images of the Plains: The Role of Human Nature in Settlement (Univ. of Nebraska Press, 1975); M.P. Lawson, The Climate of the Great American Desert (Univ. of Nebraska Press, 1974); L. Hewes, The Suitcase Farming Frontier: A Study in the Historical Geography of the Central Great Plains (Univ. of Nebraska Press, 1973); W. Kollmorgen and J. Kollmorgen, "Landscape Meteorology in the Plains Area," AAG Annals 63 (1973); T.G. Jordan, "The Origins and Distribution of Open-Range Cattle Ranching," Social Science Quarterly 53 (1972); E.C. Mather, "The American Great Plains," in J.F. Hart, ed., Regions of the United States, op. cit. (1972); J.R. Borchert, "The Dust Bowl in the 1970's," AAG Annals 61 (1971); W.M. Kollmorgen, "The Woodsman's Assaults on the Domain of the Cattleman," AAG Annals 59 (1969); J. Sims and T.F. Saarinen, "Coping with Environmental Threat: Great Plains Farmers and the Sudden Storm," AAG Annals 59 (1969); G.M. Lewis, "Regional Ideas and Reality in the Cis-Rocky Mountain West," IBG Transactions and Papers 38 (1966); T.F. Saarinen, Perception of Drought Hazard on the Great Plains

(Univ. of Chicago, Geography Research Paper 106, 1966); and L.W. Bowden, Diffusion of the Decision to Irrigate: Simulation of the Spread of a New Resource Management Practice in the Colorado Northern High Plains (Univ. of Chicago, Geography Research Paper 97, 1965).

Selected writings by geographers on the states of the interior West include C.B. Craig, Utah: A Geography (Westview, forthcoming); T.M. Griffiths and L. Rubright, Colorado: A Geography (Westview, forthcoming); R.W. Helbock and J.E. Mueller, New Mexico: A Geography (Westview, forthcoming); B. Wallach, "Sheep Ranching in the Dry Corner of Wyoming," Geographical Review 71 (1981); R.H. Brown, Wyoming: A Geography (Westview, 1980); M.L. Comeaux, Arizona: A Geography (Westview, 1980); R.L. Nostrand, "The Hispano Homeland in 1900," AAG Annals 70 (1980); R.V. Francaviglia, "The Passing Mormon Village," Landscape 22 (1978); R.H. Jackson, "Religion and Landscape in the Mormon Culture Region," in K.W. Butzer, ed., Dimensions of Human Geography (Univ. of Chicago, Geography Research Paper 186, 1978); D.C. Greer, ed., Perceptions of Utah: A Field Guide (AAG, 1977); J.D. Ives et al., "Natural Hazards in Mountain Colorado," AAG Annals 66 (1976); R.W. Travis, A Wasatch Chronicle: A Basic Bibliography of Geographic Literature on the State of Utah (Univ. of Utah, Department of Geography, Research Paper 76-2, 1976); J.M. Crowley, "Ranching in the Mountain Parks of Colorado," Geographical Review 65 (1975); J. Dupuis, "Les Montagnes Rocheuses du Colorado: Milieu Naturel et Vie Humaine," Les Cahiers d'Outremer 112 (1975); R. Symanski, "Prostitution in Nevada," AAG Annals 64 (1974), see also Commentary, Annals 65 (1975); A.W. Carlson, "Seasonal Farm Labor in the San Luis Valley," AAG Annals 63 (1973); R. Durrenberger, "The Colorado Plateau," in J.F. Hart, ed., Regions of the United States, op. cit., (1972); E.G. McIntyre, "Changing Patterns of Hopi Indian Settlement," AAG Annals 61 (1971); D.W. Meinig, Southwest: Three Peoples in Geographical Change, 1600-1970 (OUP, 1971); G.D. Weaver, "Nevada's Federal Lands," AAG Annals 59 (1969); D.B. Luten, "The Use and Misuse of a River," [the Colorado], The American West 4 (1967); S. Baker and T.J. McCleneghan, eds., An Arizona Economic and Historic Atlas (Univ. of Arizona, College of Business and Public Administration, 1966); M.J. Loeffler, "The Population Syndromes on the Colorado Piedmont," AAG Annals 55 (1965); D.W. Meinig, "The Mormon Culture Region: Strategies and Patterns in the Geography of the American West, 1847-1964," AAG Annals 55 (1965); M.J. Loeffler, "Beet-Sugar Production on the Colorado Piedmont," AAG Annals 53 (1963); M.K. Ridd, Landforms of Utah - In Proportional Relief, Map Supplement 3, AAG Annals 53 (1963); J. Hudson, Irrigation Water Use in the Utah Valley, Utah (Univ. of Chicago, Geography Research Paper 79, 1962); W.C. Calef, Private Grazing and Public Lands: Studies of the Local Management of the Taylor Grazing Act (Univ. of Chicago

Press, 1960); F.H. Thomas, The Denver and Rio Grande Western Railway: A Geographic Analysis (Northwestern Univ. Studies in Geography 4, 1960); W.W. Atwood, The Rocky Mountains (Vanguard Press, 1945): and C.D. Harris, Salt Lake City: A Regional Capital (Univ. of Chicago, Department of Geography, 1940).

Selected titles by geographers on the Pacific states include J.R. Morgan, Hawaii: A Geography (Westview, forthcoming); R.W. Pearson and D.F. Lynch, Alaska: A Geography (Westview, forthcoming); R.F. Logan, ed., AAG Field Trip Guide, 1981 [Los Angeles] (AAG, 1981); D.F. Lynch, D.W. Lantis, and R.W. Pearson, "Alaska: Land and Resource Issues," Focus 31 (1981); L.M. Cantor, "The California State Water Project: A Reassessment," Journal of Geography 79 (1980); O.E. Granger, "Climatic Variations and the California Raisin Industry," Geographical Review 70 (1980); H.J. Nelson, ed., The Los Angeles Metropolis: Readings and Syllabus ... (Bookstore, Univ. of California at Los Angeles, 6th edn., 1980); B. Wallach, "The West Side Oil Fields of California," Geographical Review 70 (1980); J. Desbarats, "Thai Migration to Los Angeles," Geographical Review 69 (1979); M.W. Donley et al., Atlas of California (Culver City CA: Pacific Book Center, 1979); R.M. Highsmith and A. Jon Kimerling, eds., Atlas of the Pacific Northwest (Oregon State Univ. Press, 6th edn., 1979); W.A. Bowen, The Willamette Valley: Migration and Settlement on the Oregon Frontier (Univ. of Washington Press, 1978); G. Dorel, "La laitue aux Etats-Unis: complémentarité des éspaces productifs et stratégie des firmes de l'agri-business," Annales de Géographie 87 (1978); C.L. Salter, San Francisco's Chinatown: How Chinese a Town? (San Francisco CA: R&E Research Associates, 1978); C.L. Yip, "A Time for Bitter Strength: The Chinese in Locke, California," Landscape 22 (1978); D.W. Lantis, R. Steiner, and A.E. Karinen, California: Land of Contrast (Kendall/Hunt, 3rd edn., 1977), a detailed geography with extremely large bibliography; J.J. Parsons, "Corporate Farming in California," Geographical Review 67 (1977); B. Rubin, "A Chronology of Architecture in Los Angeles," AAG Annals 67 (1977); J.S. Adams, ed., Contemporary Metropolitan America, op. cit. (1976), selections as follows: A.P. Andrus et al., "Seattle" (in Vol. 3), H.J. Nelson and W.A.V. Clark, "The Los Angeles Metropolitan Experience: Uniqueness, Generality, and the Goal of the Good Life" (in Vol. 4), and Jean Vance, "The Cities by San Francisco Bay" (in Vol. 2); R.W. Durrenberger and R.B. Johnson, California: Patterns on the Land (Palo Alto CA: Mayfield, 5th edn., 1976); W.G. Loy et al., Atlas of Oregon (Univ. of Oregon Press, 1976); P.R. Pryde, ed., San Diego: An Introduction to the Region (Kendall/Hunt, 1976); J.R. Shortridge, "The Collapse of Frontier Farming in Alaska," AAG Annals 66 (1976); W.A. Bowen, Mapping an American Frontier: Oregon in 1850, Map Supplement 18, AAG Annals 65 (1975); D.K. Fleming, ed., Views of Washington State (AAG, 1974); N.J.W. Thrower, Map of California Population Distribution in 1970 (California Council on Geographic Education, 1974); Department of Geography, University of Hawaii, comp., Atlas of Hawaii (Univ. Press of Hawaii, 1973), a superb atlas of the state; B.G. Vanderhill, "Perspectives on Alaskan Agriculture," Journal of Geography 72 (1973); J.E. Vance, Jr., "California and the Search for the Ideal," in J.F. Hart, ed., Regions of the United States, op. cit. (1972); D.W. Meinig, The Great Columbia Plain: A Historical Geography, 1805-1910 (Univ. of Washington Press, 1968); K.E. Francis, "Outpost Agriculture: The Case of Alaska," Geographical Review 57 (1967); R.E. Preston, "Urban Development in Southern California Between 1940 and 1965," Tijdschrift voor Economische en Sociale Geografie 58 (1967); H.P. Bailey, The Climate of Southern California (Univ. of California Press, 1966); R. Steiner, "Reserved Lands and the Supply of Space for the Southern California Metropolis," Geographical Review 56 (1966); R.W. Pease, Modoc County: A Geographic Time Continuum on the California Volcanic Tableland, Univ. of California Publications in Geography, Vol. 17 (Univ. of California Press, 1965); G.J. Fielding, "The Los Angeles Milkshed: A Study of the Political Factor in Agriculture," Geographical Review 54 (1964); H.F. Gregor, "Spatial Disharmonies in California Population Growth," Geographical Review 53 (1963); D.W. Meinig, "A Comparative Historical Geography of Two Railnets: Columbia Basin and South Australia," AAG Annals 52 (1962); R.L. Gentilcore, "Missions and Mission Lands of Alta California," AAG Annals 51 (1961); M.E. Marts and W.R.D. Sewell, "The Conflict Between Fish and Power Resources in the Pacific Northwest," AAG Annals 50 (1960); W.L. Thomas, Jr., ed., "Man, Time and Space in Southern California," Supplement, AAG Annals 49 (1959), a distinguished series of articles by specialists; D.C. Large, "Cotton in the San Joaquin Valley: A Study of Government in Agriculture," Geographical Review 47 (1957); K.H. Stone, "Populating Alaska: The United States Phase," Geographical Review 42 (1952); R. Peattie, ed., The Sierra Nevada: The Range of Light (Vanguard Press, 1947), and Peattie, ed., The Pacific Coast Ranges (Vanguard Press, 1946); J.O.M. Broek, The Santa Clara Valley, California: A Study in Landscape Changes (Utrecht: Oosthoek, 1932); and C.C. Colby, "The California Raisin Industry," AAG Annals 14 (1924). Space precludes specific listing of an imposing number of geographical writings about California in doctoral dissertations and master's theses, as well as in The California Geographer (annual), the Yearbook of the Association of Pacific Coast Geographers (annual), and many other journals. The most convenient bibliographical guide to this welter of material (as well as to numerous cognate references) is Lantis et al., California: Land of Contrast, op. cit.

Cognate references on the American West are overwhelming in number and often very high in literary quality. With its stark geographic contrasts, intractable resource questions, and dramatic history, the West has caught the imagination of many talented scholars and writers. Only a sketchy list of titles is possible here, but many of these have large bibliographies and will point the way to a practically endless number of works. This is particularly true of the individual volumes in Time-Life Books, The Old West, 24 vols. (1973-1979). Selected other titles, commencing with N.M. Hansen, The Border Economy: Regional Development in the Southwest (Univ. of Texas Press, 1981); J. McPhee, Basin and Range (Farrar, Straus, and Giroux, 1981); P.L. Fradkin, A River No More: The Colorado River and the West (Knopf, 1981); L.R. Hayes, Energy, Economic Growth, and Regionalism in the West (Univ. of New Mexico Press, 1980); D. Lavender, The Southwest (Harper and Row, 1980); L.B. Lee, "California Water Politics: Opposition to the CVP [Central Valley Project], 1944-1980," Agricultural History 54 (1980); J.F. McDermott, Jr., et al., eds., People and Cultures of Hawaii: A Psychocultural Profile (Univ. Press of Hawaii, 1980); P. Matthiessen and D. Budnick, "Battle for Big Mountain," Geo 2 (1980), a discussion of controversies concerning mineral and power development on Navajo Indian land; J.R. Milton, The Novel of the American West (Univ. of Nebraska Press, 1980); M. Parfit, Last Stand at Rosebud Creek: "Coal, Power, and People" [in Montana] (Dutton, 1980); Ron Redfern, Corridors of Time: 1,700,000,000 Years of Earth at Grand Canyon (Times Books, 1980); W.O. Spofford, Jr., et al., eds., Energy Development in the Southwest (Johns Hopkins Univ. Press for Resources for the Future, 2 vols., 1980); J.D. Weaver, Los Angeles: The Enormous Village, 1781-1981 (Capra Press, 1980); L.J. Arrington and D. Bitton, The Mormon Experience: A History of the Latter-Day Saints (Knopf, 1979); E.R. Bingham and G.A. Love, eds., Northwest Perspectives: Essays on the Culture of the Pacific Northwest (Univ. of Washington Press, 1979); T.J. Conomos, ed., San Francisco Bay, The Urbanized Estuary (American Association for the Advancement of Science, 1979); E.A. Engelbert, ed., California Water Planning and Policy: Selected Issues (Davis CA: Univ. of California Water Resources Center, 1979); J.G. Morse, Energy Resources in Colorado: Coal, Oil Shale, and Uranium (Westview, 1979); S.H. Murdock and F.L. Leistritz, Energy Development in the Western States: Impact on Rural Areas (Praeger, 1979); J.W. Reps, Cities of the American West: A History of Frontier Urban Planning (Princeton Univ. Press, 1979); J.O. Steffen, ed., The American West: New Perspectives, New Dimensions (Univ. of Oklahoma Press, 1979); H.P. Walker and D. Bufkin, Historical Atlas of Arizona (Univ. of Oklahoma Press, 1979); D. Worster, Dust Bowl: The Southern Plains in the 1930s (OUP, 1979); T.H. Creighton, The Lands of Hawaii: Their Use and Misuse (Univ. Press of Hawaii, 1978); F. Merk, History of the Westward

Movement (Knopf, 1978); D.F. Peterson and A.B. Crawford, eds., Values and Choices in the Development of the Colorado River Basin (Univ. of Arizona Press, 1978); State of California, California Water Atlas (1978), a remarkable feat of cartography, with extensive text and numerous satellite and other photos; R.B. Weeden, Alaska, Promises to Keep (Houghton Mifflin, 1978); R. Cameron, Above Los Angeles (San Francisco CA: Cameron and Co., 1977), and Cameron, Above San Francisco (Cameron and Co., 1977), books of spectacular aerial photographs; H.R. Lamar, ed., The Reader's Encyclopedia of the American West (Crowell, 1977); J. McPhee, Coming into the Country [Alaska] (Farrar, Straus and Giroux, 1977); J.M. White, The Great American Desert: The Life, History, and Landscape of the American Southwest (Allen and Unwin, 1977); R.S. Bradley, Precipitation History of the Rocky Mountain States (Westview, 1978); J. Selby, The Conquest of the American West (Rowman and Littlefield, 1976), a short history; W. Voight, Jr., Public Grazing Lands: Use and Misuse by Industry and Government (Rutgers Univ. Press, 1976 reviewed by W. Calef in Geographical Review 68, 1978); N. Hundley, Jr., Water and the West: The Colorado River Compact and the Politics of Water in the American West (Univ. of California Press, 1975); C.B. Hunt, Death Valley: Geology, Ecology, Archaeology (Univ. of California Press, 1975); J.H. Shideler, ed., "Agriculture in the Development of the Far West: A Symposium," special issue, Agricultural History 49 (January, 1975); W.A. Beck and Y.D. Haase, Historical Atlas of California (Univ. of Oklahoma Press, 1974); R.A. Billington, Westward Expansion: A History of the American Frontier (Macmillan, 4th edn., 1974); G.D. Nash, The American West in the Twentieth Century: A Short History of an Urban Oasis (Prentice-Hall, 1973); N.R. Peirce, The Great Plains States of America (Norton, 1973) and, by the same author and publisher, The Megastates of America (1972), The Pacific States of America (1972), and The Mountain States of America (1971); E. Pomeroy, The Pacific Slope: A History of California, Oregon, Washington, Idaho, Utah, and Nevada (Univ. of Washington Press, 1973); D. Seckler, ed., California Water: A Study in Resource Management (Univ. of California Press, 1971); W.A. Beck and Y.D. Haase, Historical Atlas of New Mexico (Univ. of Oklahoma Press, 1969); F.P. Farquhar, History of the Sierra Nevada (Univ. of California Press, 1966); J.R. Hastings and R.M. Turner, The Changing Mile: An Ecological Study of Vegetation Change with Time in the Lower Mile of an Arid and Semiarid Region (Univ. of Arizona Press, 1965); E.C. Jaeger, The California Deserts (Stanford Univ. Press, 4th edn., 1965); R.A. Bartlett, Great Surveys of the American West (Univ. of Oklahoma Press, 1962); G.R. Stewart, The California Trail (McGraw-Hill, 1962), a volume in the "American Trails" series; L.E. Atherton, The Cattle Kings (Indiana Univ. Press, 1961); W.H. Goetzmann, Army Exploration in the American West, 1803-

1863 (Yale Univ. Press, 1960); E.C. Jaeger, The North American Deserts (Stanford Univ. Press, 1957); P. Horgan, Great River: The Rio Grande in North American History, 2 vols. (Rinehart, 1954), see also Horgan, The Heroic Triad (Heinemann, 1974),C.F. Kraenzel, The Great Plains in Transition (Univ. of Oklahoma Press, 1955); Bernard DeVoto, The Course of Empire (Houghton Mifflin, 1952), and by the same author and publisher, Across the Wide Missouri (1947) and The Year of Decision: 1846 (1943); H.N. Smith, Virgin Land: The American West as Symbol and Myth (Random House, 1950); W.P. Webb, The Great Plains (Ginn, 1931), see also G.M. Tobin, The Making of a History: Walter Prescott Webb and "The Great Plains" (Univ. of Texas Press, 1976); J.W. Powell, Report on the Lands of the Arid Region of the United States (1878; reissued by Harvard Univ. Press, with Editor's Introduction by W. Stegner, 1962). For articles and book reviews on a miscellany of historical subjects, see The American West (bi-monthly) and Journal of the West (quarterly). Readers interested in Western landscapes are urged to view the superb photographs in Arizona Highways (monthly), the publications of the Sierra Club, and a seemingly endless outpouring of lavishly illustrated books on Western subjects by many authors and publishers. The journal American Forests (monthly) has many articles on the West.

Chapter Nine

# U.S.S.R.

C.D. Harris

The Soviet Union is the world's largest country, 22.4 million square kilometres, stretching 170° of longitude from the Baltic Sea in the heart of Europe across Asia to the Bering Strait opposite Alaska on the east. By virtue of its great longitudinal extent and high latitude it has a cool dry continental climate, vast areas of tundra, taiga (northern coniferous forest), mixed forest, steppe, and desert vegetation, and an agriculture characterized by large total production but by low yields per hectare and great variability from year to year.

Its population of 262 million in the 1979 census is less than that of China or India but somewhat more than of the United States. The Soviet Union is a multinational federal state. Politically it is composed of 15 union republics, based on ethnic groups. Russians compose just over half the population. Other major Slavic groups are the Ukrainians and Belorussians. Also along the western border are Estonians, Latvians, Lithuanians and Moldavians. In the Transcaucasus are Georgians, Armenians, and Azerbaijani. In Soviet Middle Asia peoples of Muslim tradition, speaking Turkic languages, include the Kazakhs, Kirghiz, Uzbeks, and Turkmen, and speaking Persian, the Tajiks.

Under the leadership of the Communist Party of the Soviet Union, which has been in power since 1917, the country has made great strides in developing a strong industrial base, particularly in heavy industry and machinery. In total industrial production the USSR occupies second place in the world after the United States.

Most geographical information on the Soviet Union is published in the Russian language. Serious original investigations involve the utilization of Soviet sources. There is also a substantial body of Western scholarly literature on the Soviet Union in English, French and German. This chapter is devoted to only a segment of the most accessible and valuable of these sources, primarily those in English, although certain basic sources of information in Russian are included.

Bibliographies

Two comprehensive geographical bibliographies have extensive sections on the Soviet Union. The Research Catalogue of the American Geographical Society (G.K. Hall, 1962, vol.9, pp. 6605-6803, and vol.12, pp. 8697-8820), its First Supplement Regional 1962-1971 (1972, vol.2, pp. 40-91), and its Second Supplement 1972-1976 (1978, vol.2, pp.352-361 and 493-497), has the fullest coverage of books and articles in English on the geography of the Soviet Union. Bibliographie Géographique Internationale, annual 1891-1976 and quarterly since 1977, has the best western coverage of Soviet publications on the geography of the USSR.

Guide to Geographical Bibliographies and Reference Works in Russian or on the Soviet Union, by Chauncy D. Harris (University of Chicago, Department of Geography, Research Paper no. 164, 1975), provides a detailed annotated list of some 2660 bibliographies or reference aids, mostly in Russian, for the period 1946-1973. It also includes a section on bibliographies and reference works on the Soviet Union in Western languages (pp. 353-388). It lists and annotates the principal bibliographies for each of the major systematic fields of geography, particularly geomorphology, climatology, glaciology, snow cover, permafrost, quaternary studies and paleogeography, terrestrial hydrology, oceanography, soil geography, biogeography, conservation and utilization of natural resources, economic geography, geography of population, and urban geography. Bibliographies of each of the major regional divisions of the Soviet Union are also covered. It also includes an extensive listing of the principal Russian sources on atlases (pp. 39-51), encyclopaedias (pp. 107-113), and statistical data (pp. 53-105). Guide to Russian Reference Books, volume 2, History, Auxiliary Historical Science, Ethnography, and Geography, by Karol Maichel (Hoover Institution, 1964) also provides, in section F, Geography (pp. 189-226), a detailed listing of bibliographies on Soviet geography, mostly in Russian. Soviet Geography: a bibliography, edited by Nicholas R. Rodionoff (U.S. Library of Congress, 1951), provides a comprehensive listing of older works on Soviet geography, up to 1950.

Two volumes edited by Paul L. Horecky provide good overall bibliographical coverage of studies in the various fields of the social sciences and humanities devoted to the Soviet Union: Russia and the Soviet Union: a bibliographical guide to western-language publications (University of Chicago Press, 1965), covers major works in English, French, and German, while Basic Russian Publications: an annotated bibliography on Russia and the Soviet Union (University of Chicago Press, 1962), covers works in Russian.

Retrospective bibliographies of books in English in all fields are provided by David L. Jones, Books in English on the Soviet Union 1917-73: a bibliography (Garland Publishing Co.,

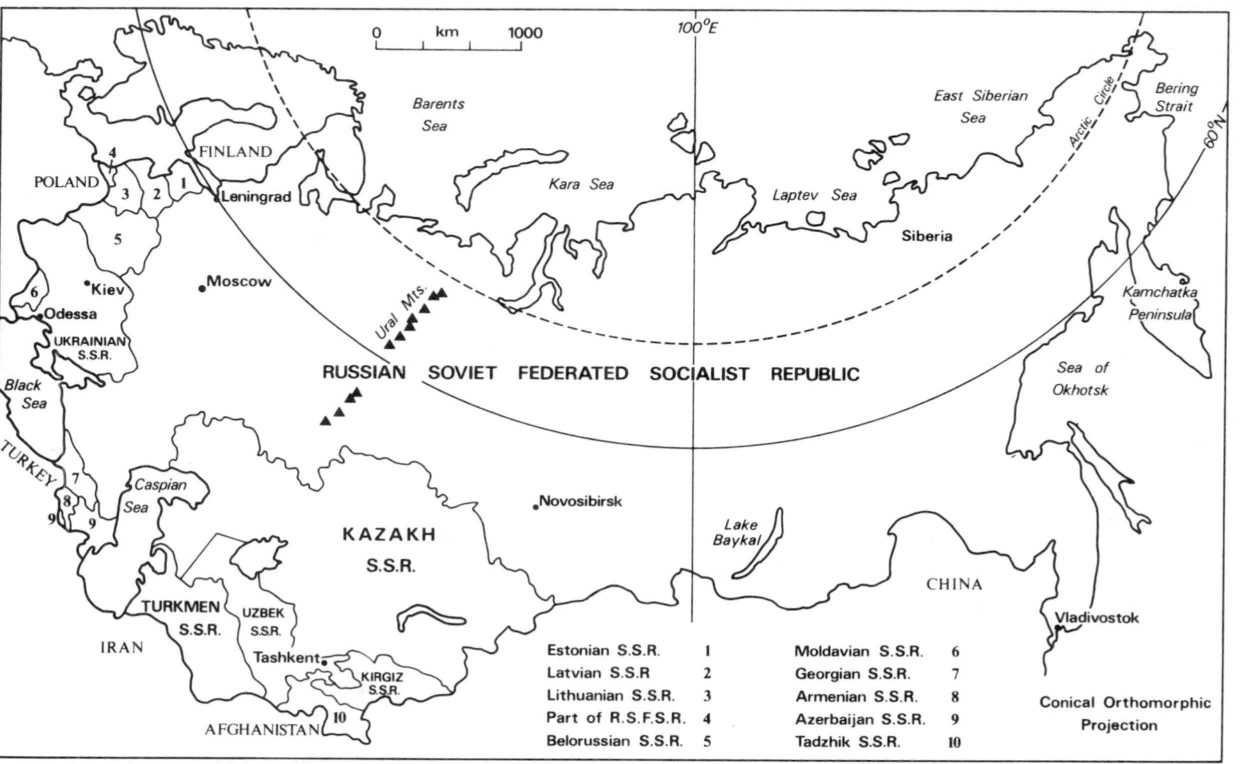


Figure 5. U.S.S.R.

1975), and Stephen M. Horak and Rosemary Neiswender, Russia, the USSR, and Eastern Europe: a bibliographic guide to English language publications, 1964-1974 (Libraries Unlimited, 1978).

Useful current bibliographies of Western work on the Soviet Union are the American Bibliography of Slavic and East European Studies (annual, 1956-1966 Indiana University, 1967, Ohio State University Press, 1968- American Association for the Advancement of Slavic Studies) and European Bibliography of Soviet and East European Slavonic Studies (University of Birmingham), which provides coverage of European publications in English, French and German.

Dissertations are listed by Jesse J. Dossick in Doctoral Research on Russia and the Soviet Union (New York University Press, 1960), Doctoral Research on Russia and the Soviet Union 1960-1975: a classified list of 3,150 American, Canadian, and British dissertations with some critical and statistical analysis (Garland Publishing Co., 1976), and later annual inventories in the Slavic Review.

The basic comprehensive current bibliography in Russian is the massive Referativnyi Zhurnal: Geografiia, monthly since 1954, issued in eleven series or in combined numbers.

Periodicals and Serials

Soviet Geography: review and translation, monthly except July and August since 1960, has continuously provided in English both translations of valuable Soviet articles in geography and news notes on the Soviet economy. For any serious work on the Soviet Union it provides a window into the extensive Russian geographical literature and conveniently assembled recent information on population, the economy, programmes, and projects. Polar Geography, quarterly since 1977, contains material on the Arctic regions of the Soviet Union. The Current Digest of the Soviet Press, weekly since 1949, provides English translations of official Soviet statements from the principal newspapers and magazines, plan goals and fulfillment, population and economic data, explanations of official and policy interpretations of principal problems.

The most important geographical periodicals in the Soviet Union itself are Izvestiia of the Academy of Sciences, series on geography, Izvestiia of the Geographical Society of the USSR, and Vestnik of Moscow State University, series 5, geography. The most interesting serial is Voprosy geografii, or Problems of Geography, sponsored by the Moscow branch of the Geographical Society of the USSR. All of these are in Russian but have supplementary tables of contents in English. The 477 geographical periodicals and serials, both current and closed, known to have been published in the Soviet Union are listed in International List of Geographical Serials, compiled by Chauncy D. Harris and Jerome D. Fellmann (3rd edn., University of Chicago, Department of Geography, Research Paper no. 193, 1980, pp. 269-342).

## Atlases

A number of small atlases in English, useful particularly for teaching, include The USSR, by G. Melvyn Howe (Hulton Educational Publications, 1972), Oxford Regional Economic Atlas: the USSR and Eastern Europe (Oxford University Press, 1956), Soviet Union in Maps, edited by Harold Fullard (G. Philip, 1965 and later), Economic Atlas of the Soviet Union, by George Kish and others (2nd edn., University of Michigan Press, 1971), and An Atlas of Soviet Affairs, by Robert C. Kingsbury and Robert N. Taaffe (Frederick A. Praeger, 1965).

Magnificent atlases published in the Soviet Union in Russian provide general-reference information, detailed thematic maps, or regional coverage. Among the most notable world atlases providing especially good information on the Soviet Union itself are Atlas Mira [Atlas of the world], edited by A.N. Baranov and others (GUGK, 1954), for general-reference hypsometric maps, available also in English as World Atlas (GUGK, 1967), Fiziko-Geograficheskii Atlas Mira [Physical-geographic atlas of the world], edited by I.P. Gerasimov and others (GUGK, 1964), for thematic maps in all fields of physical geography, Geograficheskii Atlas dlia Uchitelei Srednei Shkoly [Geographic atlas for teachers in secondary schools], edited by V.S. Apenchenko and T.M. Pavlova (3rd edn., GUGK, 1967), for distribution of ethnic groups.

The best atlas devoted solely to the Soviet Union is Atlas SSSR, edited by A.N. Baranov (2nd, edn., GUGK, 1969), with regional,physical and economic maps. Specialized thematic atlases of the Soviet Union include Atlas Razvitiia Khoziaistva i Kul'tury SSSR [Atlas of the growth of the economy and culture of the USSR], edited by A.N. Voznesenskii and others (GUGK, 1967), issued on the 50th anniversary of the October Revolution, and Atlas Sel'skogo Khoziaistva SSSR [Atlas of agriculture of the USSR], edited by A.I. Tulupnikov, (GUGK, 1960) with detailed maps of physical conditions affecting agriculture and of agricultural products.

Regional atlases of parts of the Soviet Union are numerous. Many of the union republics have their own atlases, notably the Ukrainian, Belorussian, Georgian, Armenian, Azerbaidzhan, Uzbek, and Tadzhik republics, often in two editions, one Russian, the other the language of the republic. Regional atlases not associated with political-administrative regions are devoted to the Virgin-lands (Atlas Tselinnogo Kraia, GUGK, 1964) and to Siberia east of Lake Baykal (Atlas Zabaikal'ia, GUGK, 1967). Atlases of yet smaller political units, called oblasts, number about thirty.

Statistics

Official current Soviet statistics are published in the annual statistical yearbook of the Soviet Union, Narodnoe Khoziaistvo SSR: statisticheskii ezhegodgnik (Statistika) or in the monthly periodical, Vestnik Statistiki. English-language statistical sources include USSR Facts and Figures Annual (Academic International Press, since 1977) and Ellen Mickiewicz, Handbook of the Soviet Social Science Data (Collier-Mcmillan, 1973).

For detailed analysis by regions or political administrative subdivisions of Russia or the Soviet Union one must turn to the volume in the four published censuses of 1897, 1926, 1959, or 1970: Pervaia Vseobshchaia Perepis' Naseleniia Rossiiskoi Imperii 1897 goda (1899-1905, 89 volumes), Vsesoiuznaia Perepis' Naseleniia 1926 goda (1928-1933, 56 volumes, Itogi Vsesoiuznoi Perepisi Naseleniia 1959 goda (1962-1963, 15 volumes), and Itogi Vsesoiuznoi Perepisi Naseleniia 1970 goda (1972-1974, 7 volumes). Detailed volumes for the 1979 census have not yet been published.

Textual Materials

A large number of recent books on the geography of the Soviet Union contain effective text, bibliographies, tables, maps, diagrams, and illustrations. Paul E. Lydolph in two distinctive and contrasting volumes provides on the one hand a systematic treatise on the Soviet Union as a whole in Geography of the USSR: topical analysis (Misty Valley Publishing Co., 1979) and on the other a regional treatment in Geography of the USSR (3rd edn., John Wiley, 1977). These are relatively comprehensive and up to date, with abundant statistics.

Three general surveys of the geography of the Soviet Union written by Soviet geographers have been translated into English. Soviet Union: a geographical survey, edited by S.V. Kalesnik and V.G. Pavlenko (Progress Publishers, 1976) is a systematic treatment by leading Soviet specialists and is the most comprehensive. Geography of the Soviet Union: physical background, population, economy, by V.V. Pokshishevskii (Pro gress Publishers, 1974) with a 45 page special map supplement, is particularly strong on population geography, the economy, and a regional review. Geography of the U.S.S.R.:an introductory survey, English edition edited by James S. Gregory but written by A.A. Mints (Novosti, 1975) is a brief overview.

Somewhat older general geographies of the Soviet Union still have value. The largest of these is Russian Land Soviet People: a geographical approach to the USSR, by James S. Gregory (George G. Harrap, 1968), a massive compendium on historical geography, climate, natural vegetation and agricultural zones, and general regions. The Soviet Union, by G. Melvyn Howe (Macdonald and Evans, 1968), Geography of the USSR (Macmillan, 1964), by Roy E.H. Mellor and its successor; The Soviet Union and its geographical problems by Roy E.H. Mellor

(Macmillan, 1982), *The Soviet Union: the land and its people*, by Georges Jorré (Longmans and John Wiley, 1967) and *A Geography of the Soviet Union*, by John C. Dewdney (2nd edn., Pergamon, 1971) each has its merits.

For physical geography, *Natural Regions of the USSR*, by L.S. Berg, translated from Russian (Macmillan, 1950) still provides the best detailed survey of climate, relief, vegetation, and fauna by landscape zones of the entire Soviet Union, while *Physical Geography of Asiatic Russia*, by S.P. Suslov, also translated from Russian (W.H. Freeman, 1961), contains detailed descriptions of the physical regions of the vast Asiatic part of the Soviet Union. Two good treatments of the climate of the USSR are available in English. *Climates of the USSR*, by A.A. Borisov (Oliver and Boyd, Aldine, 1965), is translated from the Russian. *Climates of the Soviet Union*, by Paul E. Lydolph (Elsevier Scientific Publishing Co., 1977), volume 17 in the World survey of climatology, provides both a regional survey and systematic treatment of the thermal factor, the moisture factor, and climate distribution. It contains numerous maps and 126 tables of climatic data.

For economic geography of the USSR one can turn to *The Soviet Union: an economic geography*, by Raymond S. Mathieson (Heinemann Educational Books or Barnes and Noble, 1975), or *A Geography of the USSR: the background to a planned economy*, by John P. Cole and F.C. German (2nd edn., Butterworths, 1970), or *Economic Geography of the USSR*, by A.N. Lavrishchev, translated from the Russian (Progress Publishers, 1969). The economic background is well treated in the *Soviet Economic System*, by Alec Nove (Allen and Unwin, 1977). Possible lines of further development are considered in *The Future of the Soviet Economy*, edited by Holland Hunter (Westview Press, 1978). Important papers by leading authorities on various aspects of the Soviet economy, presented as successive conferences, are available in *Soviet Economic Growth: conditions and perspectives*, edited by Abram Bergson (Row, Peterson, 1953), *Economic Trends in the Soviet Union*, edited by Abram Bergson and Simon Kuznets (Harvard University Press, 1963), and *The Soviet Economy Toward the Year 2000*, edited by Abram Bergson and Herbert S. Levine (in press).

A large volume edited by George J. Demko and Roland J. Fuchs, *Geographical Perspectives in the Soviet Union: a selection of readings* (Ohio State University Press, 1974) contains English translations of articles by Soviet authors on many aspects of the economic geography of the USSR, such as economic regionalization, resource management, agricultural geography, industrial geography, population geography, urban geography and historical geography.

Systematic subdivisions of economic geography are covered in a number of works, such as *Russian Agriculture: a geographic survey*, by Leslie Symons (G. Bell and Sons, 1972), and *Russian Transport: an historical and geographical survey*, by Leslie Symons and Colin White, eds., (Bell, 1975), The USSR, by John C. Dewdney (Westview Press, 1976), a volume in the series, "Studies in industrial geography," The Soviet Wood-processing Industry, by Brenton M. Barr (University of Toronto Press, 1970), Locational Factors and Locational Developments in the Soviet Chemical Industry, by Leslie Dienes (University of Chicago, Department of Geography, Research Paper No.119, 1969), and Basic Industrial Resources of the USSR, by T. Shabad Columbia University Press, 1969). The Soviet Energy System: resource use and policies, by Leslie Dienes and T. Shabad (Wiley, 1979), is a particularly valuable analysis of a timely topic. Robert W. Campbell has published three related studies: The Economics of Soviet Oil and Gas (Johns Hopkins University Press, 1968), Trends in the Soviet Oil and Gas Industry (Johns Hopkins University Press, 1976), and Soviet Energy Technologies: planning, policy, research and development (Indiana University Press, 1980). The Enigma of Soviet Petroleum: half-full or half-empty?, by Marshall I. Goldman (Allen and Unwin, 1980), provides yet another viewpoint.

Monographs on urban geography of the Soviet Union include Cities of the Soviet Union: studies in their functions, size, density and growth, by Chauncy D. Harris (Association of American Geographers, 1970), The Socialist City: spatial structure and urban policy, edited by R.A. French and F.E. Ian Hamilton (Wiley, 1979), The Soviet City: ideal and reality, by James H. Bater (Arnold, 1980), The Moscow City Region, by F.E. Ian Hamilton (Oxford University Press, 1976), Moscow: capital of the Soviet Union, a short geographical survey, by G.M. Lappo, A.G. Chikishev, and A. Bekker, translated from Russian (Progress Publishers, 1976), St. Petersburg: industrialization and change, by James H. Bater (Arnold, 1976), The City in Russian History, edited by Michael F. Hamm (University Press of Kentucky, 1976) and Patterns of Urban Growth in the Russian Empire during the Nineteenth Century, by Thomas S. Fedor (University of Chicago, Department of Geography, Research paper no.163, 1975). Population data on Soviet cities from past censuses are available in "Population of Cities of the Soviet Union, 1897, 1926, 1939, 1959, and 1967," by Chauncy D. Harris, a special issue of Soviet Geography: review and translation, May 1970.

An Historical Geography of Russia, by William H. Parker (University of London Press, 1968) provides pictures of Russia in various past periods. More specialized but valuable studies are The Russian Colonization of Kazakhstan: 1896-1916, by George J. Demko (Indiana University, 1969), and Feeding the Russian Fur Trade: Provisionment of the Okhotsk Seaboard and the Kamchatka Peninsula 1639-1856 (University of Wisconsin Press, 1969).

Two volumes provide extensive analyses of the distribution of population by regions over the period since the 1897 census. Nationality and Population Change in Russia and the USSR: an evaluation of census data, 1897-1970, by R.A. Lewis, R.H. Rowland, and R.S. Clem (Praeger, 1976), studies regional changes in ethnic groups of the multi-national Soviet Union over successive censuses, 1897, 1926, 1959 and 1970. A companion volume, Population Redistribution in the USSR: its impact on society, 1897-1977, by R.A. Lewis and R.H. Rowland (Praeger, 1979), focuses on changes in regional patterns of urbanization, urban growth, city size, urban regions, and rural population over 80 years. Valuable analyses of the 1979 census are contained in Soviet Census 1979, (GeoJournal, 1980). Papers on Soviet population and policy are contained in Soviet Population Policy: conflicts and constraints, edited by Helen Desfosses (Pergamon Press, 1981).

Ethnic diversity of the Soviet Union has aroused great interest in recent years. The Handbook of Major Soviet Nationalities, edited by Zev Katz and others (Collier Macmillan, 1975), provides a good basic source of data and information. Collections representing a wide variety of viewpoints and disciplines are now available: Soviet Nationality Policies and Practices, edited by J.R. Azrael (Praeger, 1978), Ethnic Minorities in the Soviet Union, edited by E. Goldhagen (Praeger, 1968), Soviet Asian Ethnic Frontiers, edited by W.O. McCagg, Jr., and B.D. Silver (Pergamon Press, 1979), and several works edited by Edward Allworth, Soviet Nationality Problems (Columbia University Press, 1971). The Nationality Question in Central Asia (Praeger, 1973), Nationality Group Survival in Multi-Ethnic States: shifting support patterns in the Soviet Baltic region (Praeger, 1977), and Ethnic Russia in the USSR: the dilemma of dominance (Pergamon Press, 1980).

The last decade has witnessed a burgeoning interest in environmental problems and resource management in the Soviet Union. First is a collection of studies translated from the Russian, Natural Resources of the Soviet Union, edited by I.P. Gerasimov, D.L. Armand, and K.M. Yefron (Freeman, 1971). Conservation practices and policy have been examined by Philip R. Pryde in his book Conservation in the Soviet Union (Cambridge University Press, 1972). A wide-ranging series of papers on resources and environment has been edited by W.A. Jackson, Soviet Resource Management and the Environment (American Association for the Advancement of Slavic Studies, 1978). Pollution and similar problems have recently received much attention. F.B. Singleton has edited a volume, Environmental Misuse in the Soviet Union (Praeger, 1976), volume 5 of the papers from the First International Slavic Conference, Alberta, 1974, with case studies on air pollution by V.Mote, water pollution by C. Zumbrunnen, and surface-mined land by P.R. Pryde. An over-all view is presented by M.I. Goldman in Environmental Pollution in the Soviet Union: the spoils of progress (MIT, 1972).

Information Sources

Regional studies of parts of the Soviet Union in Western languages are not very abundant. Mention may be made, however, of Gateway to Siberian Resources, by T. Shabad and V.L. Mote (Wiley, 1977), Siberia Today and Tomorrow: a study of economic resources, problems, and achievements, by Violet Conolly (Collins, 1975), The Circumpolar North: a political and economic geography of the Arctic and Sub-Arctic, by T. Armstrong, G. Rogers, and G. Rowley (Methuen, 1978), Russian Settlement in the North, by T. Armstrong (Cambridge University Press, 1964), The Soviet Far East, by E. Kirby (Macmillan, 1971), The Soviet Far East: a survey of its physical and economic geography, by E. Thiel (Praeger, 1957), and Central Asia: a century of Russian rule, edited by E. Allworth (Columbia University Press, 1967), with chapters on the land, population, agricultural development, and industrialization by I. Matley. Russo-Chinese Borderlands: zone of peaceful contact or potential conflict, by W.A.D. Jackson (Van Nostrand, 1962), is a study in political geography. A New Soviet Heartland?, by D. J.M. Hooson (Van Nostrand, 1964), presents a brief examination of the regions of the deep interior of the Soviet Union from the Volga River to Lake Baykal. Rail Transportation and the Economic Development of Soviet Central Asia, by R.N. Taaffe (University of Chicago, Department of Geography, Research Paper no.64, 1960), and Transport in Western Siberia: Tsarist and Soviet Development, by R.N. North (University of British Columbia Press, 1979), analyse the role of transportation in regional specialization and development.

Encyclopaedias

The 30-volume Great Soviet Encyclopaedia is being translated into English (Macmillan and Collier-Macmillan, 1973- ). It provides articles on places, regions, and systematic fields in geography and other disciplines. Although still in progress, 26 of the volumes had been published by early 1981. It is a translation of Bol'shaia Sovetskaia Entsiklopediia (3rd edn., Sovetskaia Entsiklopeiia, 1970-1978). The New Encyclopaedia Britannica (15th ed., 1974) contains many articles on the Soviet Union. Of special note are those by Soviet physical geographers. A specialized 5 volume geographical encyclopaedia in Russian which has many valuable articles on all phases of the geography of the Soviet Union is Kratkaia Geograficheskaia Entsiklopediia (Sovetskaia Entsiklopediia, 1960-1965).

TOOLS FOR THE GEOGRAPHER

Chapter Ten

# MAPS, ATLASES AND GAZETTEERS

G.R.P. Lawrence

## Introduction

It has never been seriously disputed that the map is one of the geographer's principal tools both for source material and also as a means of presenting data and theories. Neither is it contended that the Geographical Sciences are essentially 'spatial'. Conversely, the printed or published map record is often taken as definite, precise, beyond question and utterly correct. The serious student of cartography will know that this is not so. The map is a substitute for reality, it is a model and is only as good as the source material available and it is also restricted by the limitations of cartographic techniques. These constraints stem from the fact that the map is a representation to scale of the features it purports to show; this information is therefore generalised and symbolised.

Cartographical reliability may be the hallmark of official map series such as those produced by organisations like the Ordnance Survey of Great Britain, but there are instances of cartography being used in a subjective and even misleading way in order to present a biased viewpoint, e.g. as propaganda maps in time of war or in order to emphasise town or tourist features at the expense of surrounding areas. Only the map, however, can adequately portray the various facets of the earth's surface; although the value of the air photograph can be gauged from the details of Chapter 11.

Maps can be used to portray information relating to any of the many branches of Geography, ranging from relief and landforms through soils and vegetation to the data of human activities and population distribution as well as including important but abstract information such as boundaries.

## Map Types

The only true representation of the earth's surface is a globe. Anything else on a flat sheet must be obtained by means of a map projection. For further information on this subject see:

P.W. McDonnell, Introduction to Map Projections, (Dekker, 1979)
H.S. Roblin, Map Projections, (Arnold, 1969)
J.F. Steers, An Introduction to the Study of Map Projections, (15th edition, Univ. of London Press, 1978)

Where maps are at a large scale and cover a very small area of the earth's surface the projection becomes of little significance.

For maps of large areas and therefore at small scales such as those of continents, hemispheres or the entire world, different projections may be used depending on the purpose of the map. All reliable mapping by official or reputable organisations will state both the projection used and also give reference information such as the map graticule (lines of latitude and longitude) or a map grid reference (referring to a system of numbered squares on the map) and will also state the scale. (It should be stressed here that a map without a scale is valueless). On small scale maps the scale will only be correct along particular lines, or a scale may be quoted which refers to that of the globe or reference spheroid on which the projection is based. Details of scale, projection, etc. may be placed either within the border of the map or in the margins.

The map itself may be unique in that there is only one copy, in manuscript form, but it is more likely to be one of many reproduced by a printing process of some kind. Early maps, prior to the introduction of printing, of course only exist in the single manuscript editions (see reference on the History of Cartography, below).

Maps may be issued as single sheets, parts of a series, part of a collection in atlas form (see later), as illustrations to articles or in books, or as large format wall maps for display or consultation. They may be available as "flat" sheets or folded, and if the latter, may be in covers or "cased". Indexes or gazetteers may be issued with individual maps or as separate publications to accompany a whole map series. The arrangement of maps in a series may be described or illustrated by a diagram of sheet lines or notes on adjacent sheets either in the margins or, as is the case with the sheets of the Ordnance Survey 1/25,000 series, in a diagram on the reverse of the map. Alternatively, descriptive literature in single sheet or booklet form may be available to explain these matters.

The original map document, or "archive", remains in the hands of the map-making organisation and is used to create other maps, normally at smaller scales and therefore providing less information because of the limitation of the scale. In order to make up-to-date survey information available to map users the archive data can be reproduced in various ways :

a) in conventional printed map form

b) in digital form on tape or disc

c) from digital data fed to a processor which will then draw a map on a display screen and/or a plotter

d) as printed enlargements from small transparencies (generally printed as a diazo sheet)

Methods (b) and (c) may be termed computer maps whilst (d) are known as aperture cards. In the case of the Ordnance Survey this flow of current information at a large scale (1/1250 or 1/2500) is termed SUSI (Supply of Unpublished Survey Information) and SIM (Survey Information on Microfilm). The relevant aperture cards are held by official Ordnance Survey agents throughout Britain.

Maps can be divided into two main categories - Topographic and Thematic. Topographic maps depict all the features of the appropriate area that can be included at the chosen scale whilst the latter are essentially selective. Thematic maps are not necessarily single-purpose, as compared with the multi-purpose topographic, and they may be drawn to represent an almost limitless variety of special features by anyone wishing to depict spatial distributions. A further category of map form within the broad thematic group is the <u>cartogram</u> (note that this term is also used to mean a symbol created for mapping purposes). These are maps in which spatial relationships have been simplified and are best exemplified by transport route maps, for example the well known maps of the London Underground network and the Subway system of New York. Hypothetical patterns in research studies may also lead to maps of this type, e.g. urban hierarchies, zones of influence.

Both topographic and thematic maps may be produced in single or multiple colours and although there are various colour conventions in cartography it is not always necessary to follow these. (Table 2 indicates some of these conventions). In order to project a particular message the designer/cartographer may well use methods which do not conform to normal usage. There are also many conventional symbols in current use on both topographic and thematic maps and these should of course be described in a "legend" or "key" to the map. In some cases only a summary legend may appear on the map with a fuller version supplied on a separate sheet or card. Basic information on map types is covered in the following books:

G.C. Dickinson, <u>Maps and Air Photographs</u>, (Arnold, 1969)

G.R.P. Lawrence, <u>Cartographic Methods</u>, (Methuen, 1979)

P.C. Muehrcke, <u>Map Use</u>, (J.P. Publications,1978)

Table 2
Colour Usage for Different Phenomena on Maps

1 BLACK for details of settlement and ground-plan lines, point information, lettering and topographic features such as rock outcrops; also for boundaries

2 Strong BLUE for drainage network

3 Reddish BROWN for contours and also possible for rock outcrops (GREY can also be used)

4 Strong GREEN for vegetation, tree and woodland symbols

5 GREY for shading or hatching lines of built-up areas, buildings

6 YELLOW for any distributional tones required or additional boundary lines (as zone alongside lines)

7 Dark BROWN for additional detail

8 Strong RED for communications network

9 Light BLUE for areas of water, seas, lakes

10 Light RED available for higher relief shading

11 Light GREY for relief shading

The first three can be considered as the major colours of the linework on the topographic map.

These colours will show through those of the next group (4,5 and 6) which serve to complement 1,2 and 3. The remainder are supplementary colours.

Note that 1,2,3,4,7,8 are strong colours, to be used for lines and symbols in the main.

5,6,9,10 and 11 are light colours, to be used chiefly for shadings but possible for linework.

6,9 and 10 can be used together and in various strengths to produce a range of shadings.

Blue and blue mixtures are generally regarded as *cold* colours; use these therefore in arctic and alpine areas, e.g. icefields, higher relief.

Red and red mixtures are generally regarded as *warm* colours.

---

Source :Lawrence Cartographic Methods

A.H. Robinson, R.D. Sale, and J. Morrison, Elements of Cartography, (Wiley, 1978)

An indication of the range of thematic map types is given in Table 3 which attempts to categorise them and which may form the basis of a classification system.

## About Maps - Past and Present

Map-making in its widest sense embraces all stages of map preparation from surveying and data collection to the draughting and printing (or other form of reproduction) of the final map. Common usage in many countries, however, confines map-making to the later stages of the full process, i.e. to that of drawing and reproduction.

In addition to the books listed above, further general information on maps, their making and their use, can be obtained from the following:

J.B. Harley, Ordnance Survey Maps - a descriptive manual, (O.S., 1975)

A. Hodgkiss, Maps for Books and Theses, (David & Charles, 1970)

War Office, Manual of Military Engineering, Volume XIII, (Ministry of Defence, 1967-71)

Map-using includes relatively straight-forward tasks like the location of points and places (e.g. using grids and finding streets on a town plan) as well as more sophisticated activities such as the recognition and delimitation of landforms, settlement types and the identification of possible geographical inter-relationships. These last activities are probably better understood as map-reading.

Additional information on basic map-reading skills, contour interpretation and generally learning the basics of relating the topographic map to the outdoor landscape, will be found in any of a large number of textbooks prepared for the middle school, such as A.P. Fullager and H.E. Virgo's Map-reading and Local Studies in Colour, (English U.P., 1975).

The historical development of the subject is a further aspect which can appeal to the student of Cartography. The study covers a wide time span, from early geographers' views of the size and shape of the earth in early times to the development of official mapping organisations in the last century. The subject matter and its portrayal are also inextricably linked with man's developing knowledge of the world and also with the development of artistry and technology, particularly printing. An important new book in this field is Mapmakers by J.N. Wilford (Vintage Books, 1982). The reader is also referred to:

Table 3

Categories of Thematic Maps

| | | |
|---|---|---|
| RELIEF representation, and the PHYSICAL landscape | e.g. | depiction of landforms, drainage, soils, geology, hydrology |
| METEOROLOGICAL and CLIMATIC | e.g. | Weather maps, synoptic charts, rainfall, temperature, pressure, winds |
| HYDROGRAPHIC | e.g. | Depths of water, navigation aids, harbour plans |
| AERONAUTIC | e.g. | Flight paths, airports |
| LAND USE and VEGETATION | e.g. | Agriculture, land classification, land capability, farms, types of farming, woodland, moorland, types of vegetation, forestry |
| URBAN, SETTLEMENT and POPULATION | e.g. | Urban areas, town types, spheres of influence, population distribution, population density, migrations, occupations, households, service industries |
| HISTORICAL and ARCHAEOLOGICAL | e.g. | Former landscapes, settlement growth, reconstructions of settlement forms, development of communications, etc. |
| COMMUNICATIONS | e.g. | Routes by rail, road and water, airlines, trade, (showing either classification or volume) |

| | |
|---|---|
| ADMINISTRATION | e.g. International, national boundaries at various levels, town functions |
| RECREATION | e.g. Tourism, parkland, sites of local/national interest, access |
| PLANNING | e.g. Types of housing, retail areas, shopping patterns, state of national, regional, local plans, proposed developments |
| RESOURCES | e.g. Minerals, extractive industry, heavy industry, factories, employment |

The above list cannot pretend to be exhaustive but is merely given to indicate some of the possible map types likely to be encountered in a classification of flat sheets or in atlas form for particular regions or countries.

---

Source:Lawrence _Cartographic Methods_

Lloyd A. Brown, Story of Maps, (Dover Press,1980)

G.R. Crone, Maps and their Makers (4th edn., Hutchinson, 1970)

R.A. Skelton, History of Cartography, (Watts,1964)

Those with a continuing interest in historical cartography will also find much to interest them in the scholarly journal, Imago Mundi (published irregularly but containing well illustrated and pertinent articles in this subject). Those who are tempted to study this area further and collect old maps will also be interested in The Map Collector, a journal for the antique map collector edited by R.V. Tooley, himself an author of textbooks dealing with old maps - recent titles being Maps and Mapmakers,(Batsford,1978) and Tooley's Dictionary of Mapmakers, (Map Collectors Pr.,1979). A small biannual journal which includes articles of a general nature on many aspects of map collecting is The Map Reader. It is published by the Southern African Map Collectors Association which is based at the University of the Witwatersrand in Johannesburg.

For detailed information on the availability of maps for all parts of the Earth's surface, the reader is recommended to consult Modern Maps and Atlases by C.B. Muriel (Lock,1969) or its successor by the same author, Geography and Cartography, a Reference Handbook, (Bingley,1976). A work which gives a full and fairly up-to-date record of map coverage generally is K.L. Winch's International Maps and Atlases in Print, (2nd edition, Bowker, 1976). Alternatively, one of the following publications of the GeoCenter Internationales Landkartenhaus may be consulted:

GeoKatalog, Band 1, an annual publication which contains details of official and commercial tourist maps, plans, guides,atlases and globes.

GeoKatalog, Band 2, which is a continuous publication, looseleaf, arranged by countries and which as well as containing much of the material that will appear in Band 1 also contains index maps of official publications. Unfortunately not all countries are covered.

Kartenbrief, a periodical supplement received by subscribers to Band 2, it appears quarterly and lists new maps and therefore keeps the main volume abreast of new developments.

The address of the GeoCenter Internationales Landkartenhaus is: Postfach 800830, D.7000 Stuttgart 80, Germany F.D.R.

## Mapmakers, Publishers and their Agents

National mapping bodies exist in most countries. A list of some of these organisations is given in Table 4, and further details may be found in GeoKatalog 2 (referred to above). Maps produced by such bodies will usually conform to a national specification and their sheet lines will form a national series.

The official mapping organisation for the United Kingdom is the Ordnance Survey based in Southampton (Romsey Road, Maybush), and for France it is the Institut Géographique National in Paris. In West Germany (the Federal Republic) each Land (or Province) carries out its own surveying and map publishing, however The Institut für Angewandt Geodesie in Frankfurt acts as something of a clearing house.

There is also a private and commercial sector in map-making and publishing in most countries. Unlike the official mapmakers these private organisations do not always conform to national specifications. As with their official counterparts, however, details of their work appears in GeoKatalog 2, albeit rather selectively. In the U.K. some of the main commercial map publishers are: Bartholomew, Geographica, O.U.P., Philip's, the Automobile Association and Readers Digest. Certain leading American map publishers are: Rand McNally, Denoyet-Geppert, Hagstrom and the American Map Company.

## Sources of Information on Current Map Publications

An additional role played by commercial publishers in some instances is that of agent for the various national organisations from other parts of the world. A selection of the major maps published by the organisation or organisations for which the firm acts as agent will often appear in that firm's catalogue. Persons wishing to acquire maps may do so through normal retailers. However certain retailers have specialised in the map trade and are themselves agents for the various national mapping organisations. Stanfords of Long Acre in London are probably the best known example of this in the U.K. If retailers and agents fail to satisfy your requirements, as a last resort it is possible to get in touch with the various organisations themselves.

Information on map publications, maps in print, etc. are provided by publishers'catalogues. Official mapping organisations usually issue regular notification of their publications, for example, the new publications of the British Ordnance Survey are listed monthly to supplement the basic annual Map Catalogue of the Survey. Similarly, British Geological Survey maps are noted in a quarterly list. These publications also appear in the Ordnance Survey's annual Map Catalogue.

Table 4

Selected List of Official Mapping Organisations

| | |
|---|---|
| AUSTRALIA | Division of National Mapping<br>Department of Minerals and Energy, Canberra |
| AUSTRIA | Bundesamt für Eich-und-Vermessungswesen, Vienna |
| BELGIUM | Institut Géographique Militaire, Brussels |
| CANADA | Department of Energy,Mines and Resources, Ottawa |
| EIRE | Ordnance Survey, Dublin |
| FRANCE | Institut Géographique National, Paris |
| GERMANY, West | Institut für Angewandt Geodesie, Frankfurt |
| ITALY | Instituto Geografico Militare, Florence |
| JAPAN | Geographical Survey Institute,<br>Ministry of Construction, Tokyo |
| NETHERLANDS | Topografische Dienst, Delft |
| NEW ZEALAND | Lands and Survey Department, Wellington |
| PORTUGAL | Instituto Geografico e Cadastral, Lisbon |
| POLAND | Glówny Urzad Geodezii Kartografii, Warsaw |
| SPAIN | Servicio Geografico del Ejercito, Madrid |
| SOUTH AFRICA | Trigonometrical Survey Office, Pretoria |
| SWITZERLAND | Eidgenössische Topographische, Berne |
| UNITED KINGDOM | Ordnance Survey, Southampton |
| U.S.A. | United States Geological Survey, Reston, Virginia |

Britain's Department of the Environment's maps are listed in their Map Library's _Classified Accessions List_, formerly quarterly, this is now _an irregular publication_.

Other listings, such as those in journals, tend to be selective, but the following should be noted: _Recent Maps and Atlases_ in _The Cartographic Journal_ is compiled from recent accessions to the Bodleian Library, Oxford. The Bodleian is one of the "copyright" libraries, as is the British Library, London, and the Cambridge University Library and so receives copies of all publications in the U.K., including maps. _The Cartographic Journal_ list also notes major overseas maps received at Oxford and publishers' catalogues. Until 1981 the Royal Geographical Society in London published an annotated list of new additions to its Library and Map Room entitled _New Geographical Literature and Maps_. Now that this has sadly ceased publication, reference to Royal Geographical Society Map Room additions are hopefully also to appear in this important publication. _The American Cartographer_ is also worthy of mention in the context of provision of information on new maps, etc.

Other sources of information are the various geographical periodicals, such as the _Geographical Journal_, which includes a section "Cartographical Survey". Although the emphasis may be rather more on books and articles in journals, there are also references to maps in _GeoAbstracts (Series D)_ and in the extensive American publication, _Current Geographical Publications_.

A bi-monthly information fact sheet entitled _Cartactual_, which can be of use to the map researcher, is produced by the Hungarian National Office of Lands and Mapping and this has a more specific cartographic supplement, _Cartinform_. This work is particularly useful for information on atlases and, as might be expected, also useful for guidance on East European publications in mapping. Similarly, the annual publication, _Geographical Digest_, produced by George Philip & Sons, London, contains useful up-to-date information on maps as well as tables of data and other statistics. This group of publications is a reflection of the need for map publishers to be in touch with current geographical topics and they include information on such subjects as changes in administrative boundaries, new settlements, civil engineering developments, name changes, transport and communications activities, etc.

## Atlases

Although maps and atlases are broadly grouped together for many purposes and are generally kept fairly close together in map libraries, there are important differences in the characteristics of maps in atlas form. The atlas is more clearly a reference source but they come in many different forms.

The basic definition of an atlas, given in a Glossary of Technical Terms in Cartography (Royal Society, 1966), is "a collection of maps designed to be kept together in a bound or loose format". The Glossary also distinguishes between national and regional atlases and the definition may be strengthened by remarking that it is normal for the maps to have some logical sequence. Further forms of atlas are those prepared on either a broad or world scale and those which possess a particular theme. Again, atlases may be designed for library use, as is the case with The Times Atlas of the World, Comprehensive Edition (Times Books, 1980), or they may be intended for one of many levels in education where multiple copies will be required for class or examination use. In both cases these are world atlases and generally have a preponderance of topographic maps together with some thematic and a small number of large scale town maps or regional maps.

Regional and national atlases have in common the fact that they deal with one specific area of the world and often have a very limited range of base maps upon which different thematic information is published. A fairly typical example illustrating these points is the New Zealand Atlas published in Wellington (N.Z. Government Printer, 1976). This splendid work contains 203 maps of New Zealand dealing with every aspect of the country's life as well as a large number of coloured photographs, and many pages of informative text.

Thematic atlases are frequently produced to cover a narrow areal range: for example, one may refer to An Agricultural Atlas of England and Wales, (2nd edition, Faber, 1974), An Agroclimatological Atlas of the Northern States of Nigeria (Ahmadu Bello U.P., 1972), Atlas of Ancient Archaeology (Heinemann, 1974) or Character of a Conurbation (Hodder & Stoughton, 1971). The last-named exemplifies the single-colour computer atlas and contains a number of maps of Birmingham and the West Midlands showing various population parameters as disclosed in an analysis of the 1966 Census. A recent atlas in which computer assistance featured for both the analysis of the data and the technical preparation of the reproduction material is People in Britain - a Census Atlas, (HMSO, 1980). This is a multi-colour production in A4 format and its production is described in some detail in an article in The Cartographic Journal for June 1980.

A recently published thematic atlas that deserves special attention is Michael Kidron & Ronald Segal's The State of the World Atlas (Heinemann, 1981). The work consists of 65 maps displaying innovatory cartographic techniques and graphic design and illustrating many of the major forces that are shaping world geography in the 1980's. Two typical maps are those entitled 'Food Power' and 'Bullets and Blackboards'. The former attempts to show the world from the point of view of food trade deficit and surplus. The latter shows the world on the basis of a ratio of soldiers to school teachers - ranging from the countries of the Middle East with more than 250 soldiers per 100 teachers to countries such as Canada and India that have more than 250 teachers per 100 soldiers.

## Gazetteers

The location of places on maps demands a satisfactory gazetteer and this in turn requires a clear and precise technique by which the placenames may be found on the maps themselves. There is no substitute for the conventional method of latitude and longitude as a universal location system. When dealing with some map series, however, care should be taken to read the International figures in respect of longitude (i.e. in degrees and with reference to the International, Greenwich, datum) and not to mistake figures quoted in grads or referring to a local datum, e.g. the meridian of the national capital. Some topographic maps of France provide a good illustration of such reference schemes in use.

On maps at larger scales,and particularly on those which form part of a national series,inter-related with other scales it is usual to find a grid system, often related to the map sheet size and boundaries. The National Grid Reference System as used by the Ordnance Survey is a familiar British example but others are the Universal Transverse Mercator system, and Georef. Further information regarding these systems will be obtained in G.R.P. Lawrence, _Cartographic Methods_ (Methuen, 1979) or G.C. Dickinson,_Maps and Air Photographs_ (Edward Arnold, 1969).

World gazetteers are a necessary index feature of all world atlases and for many purposes the gazetteer/index to _The Times Atlas of the World Comprehensive Edition_ (Times Books, 1980) will prove adequate. National map-making organisations also produce some form of gazetteers to assist place location in their country, as do tourist and motoring organisations (these last named may often be derived from the national ones and use a modified form of the national reference-giving system). The Ordnance Survey Gazetteer of Great Britain is based on the place names shown on the Fifth Edition of the 1/250,000 map and is therefore only as complete as that particular map series.

In order that a gazetteer may be used efficiently there must be agreement in such fundamental information as the spelling of place names and, for example, with Hebrew, Arabic or Chinese names, their transliteration or romanisation. This problem of transliteration is aggravated by the fact that different transliteration schemes have sometimes been in vogue at different times in history. An example of this is the transliteration of Chinese. Up until quite recently the Wade-Giles Scheme was generally used. Now, however, the Chinese Government has introduced its own scheme - Pin-Yin. Kuang-chou (Wade-Giles) has now become Guangzhou (Pin-Yin),

Pei-ching is now Beijing. This latter example illustrates a further complication - the town is more often referred to in the West by its anglicised name, Peking. However, it is generally the practice that the spelling used on a map will be that of the country mapped. Where alternatives are possible these may also be shown in parentheses, if the scale of the map allows. There are special problems with countries or regions that are multi-lingual e.g. Belgium, and also with towns or places where common usage has produced a further form used widely outside the national borders. Again, Belgium provides an example with Brussels, Bruxelles, Brussel, (the Anglicised, French and Flemish forms respectively). Local usage may also be at variance with the names given on large scale topographic maps although major efforts are usually made by map surveyors in the field to reduce such inconsistencies.

Another problem with place-names is that they have tended to change in many parts of the world over the years for political reasons. A. Room's valuable work Place-name Changes since 1900: a World Gazetteer (Methuen,1979) is a great help in this context as is the book by Julie Wilcocks, Countries and Islands of the World: a Guide to Nomenclature, (Bingley, 1981). Authorities which may be consulted by the mapmaker or map user for advice on place names are the Lists of the Place Names, published by the Permanent Committee on Geographical Names (P.C.G.N.) in London, and the Gazetteers of the U.S. Board on Geographic Names.

A useful world wide listing of gazetteers is given in C.M. Winchell's Guide to Reference Books and its three supplements (American Library Association,1967-1972). These important reference books are available in the reference division of most public or university libraries.

## Map and Atlas Collections

The World Directory of Map Collections, compiled by the International Federation of Library Associations (IFLA) and edited by Walter W. Ristow (Verlag Dokumentation, 1976) is an invaluable starting point for anyone wishing to learn where maps may be referred to - whether they are in print or not. The volume lists over three hundred collections in numerous different countries - providing information about the size of their holdings, their annual intake and their accessibility to the general user. 56 collections in the U.S. are referred to and 27 in the U.K. Of the considerable number in London that are listed three are especially worthy of mention. The British Library (Reference Division) in Bloomsbury was said to hold 1,250,000 map sheets, 12,250 manuscript maps, 20,000 atlases and 50 globes. 25,000 maps are said to be added annually. The Map Room of the Royal Geographical Society in Kensington is described as holding 535,000 maps, 350 manuscript maps, 12,300 atlases and 35 globes. 8,800 maps and 50 atlases are accessioned annually. And thirdly, the Map collections of the library of the University of London at Senate House in Bloomsbury are said to have 42,000 maps and to add 1,500 new ones every year.

Under this head the reader's attention should also be drawn to 'Map Libraries', David Ferro's contribution to British Library and Information Work 1976-80 (Library Association, 1982). The work provides an admirable summary of developments in map libraries, especially in Britain but to a certain extent in North America as well, during the five years from 1976 to 1980. Initially the author discusses major contributions to the literature of map librarianship in this period and the book by Dr Walter Ristow, formerly map librarian at the Library of Congress entitled The Emergence of Maps in Libraries (Mansell, 1980) stands out. Subsequently in discussing significant acquisitions to map libraries, Aberdeen University's obtaining the U.S. Geological Survey's topographical coverage of the U.S. at 1:24,000 (around 50,000 sheets) is an interesting development. Library Trends (University of Illinois Graduate School of Library and Information Sciences) for the winter of 1981 also contains a wealth of valuable information. This number of the world famous journal is edited by Mary Lynette Larsgaard, map librarian at the Colorado School of Mines, and is dedicated entirely to Map Librarianship and Map Collections. Following discussion of such topics as topographic map acquisition and microcartography there is an important survey of the 70 largest map collections in the USA and also discussion of similar collections in Australia and New Zealand.

## Appendix LIST OF CURRENT CARTOGRAPHIC SERIALS

| Title | Country | Frequency | Language | Brief Notes |
|---|---|---|---|---|
| CARTOGRAPHY (Australian Institute of Cartographers). | Australia 1-(1954-) | 2 per annum | English | Papers on the science and art of cartography. News of events and people in cartography, particularly in Australia but also internationally. Reviews. Abstracts of articles on cartography in other publications. Literature received. (v 11 no 1 (Mr 1979)). |
| ASSOCIATION OF CANADIAN MAP LIBRARIES. ASSOCIATION DES CARTOTHEQUES CANADIENNES. BULLETIN. | Canada 1- (1968-) | 3 per annum | English/ French | Articles, annual conferences, reviews, committee reports, and notices and communications on maps, cartobibliography, map librarianship, and cartography. |
| CANADIAN CARTOGRAPHER (Canadian Cartographic Association). | Canada 1- (1964-) | 2 per annum | English | Articles on geographical cartography. Book reviews. Cartographic commentary. Recent cartographic literature. Inter national board of contributing editors. |
| CARTOGRAPHICA | Canada 1- (1971-) | 2 per annum | English | Monographs on cartography, either a single major work or a collection of research papers relevant to a principal theme. (no 24 (1979)). |

| | | | | |
|---|---|---|---|---|
| BIBLIOGRAPHIE GEOGRAPHIQUE INTERNATIONALE. INTERNATIONAL GEOGRAPHICAL BIBLIOGRAPHY (France. Centre National de la Recherche Scientifique. Laboratoire d'Information et de Documentation en Geographie). | France 1- (1891-1976) | 4 per annum plus index | French | In many ways the most convenient, best balanced and most international of the current geographical bibliographies. No other work so fully and intelligently records the corpus of geographic literature over the last 90 years. International group of collaborators. Covers all fields of geography. Each issue contains a list of periodicals analysed, subject index, place index, and author index; these indexes cumulate into annual volumes, which are very useful. (v 84 (1979) no 4 and index vol). |
| COMITE FRANCAIS DE CARTOGRAPHIE. BULLETIN. | France 1- (Mr 1958-) | 4 per annum | French | Articles on cartography. Reports of national and international meetings. Reports of commission activities. National and international information. Bibliography. Editorials. (no 80 (Jn 1979)). |

| Title | Country | Frequency | Language | Brief Notes |
|---|---|---|---|---|
| BIBLIOGRAPHIA CARTOGRAPHICA Internationale Dokumentation des kartographischen Schrifttums. International documentation of cartographical literature. Documentation internationale de la litterature cartographique. | Germany 1- (1974-) | Annual | German/ French/ English | Bibliography of cartography including map collections, documentation, publications, history of cartography, institutions and organizations, theoretical cartography, technology, topographic and landscape cartography, thematic maps and cartograms, atlas cartography, uses of maps, special purpose maps, relief representation, and globes. List of periodicals analysed. (5 (1978)). |
| INTERNATIONAL YEARBOOK OF CARTOGRAPHY. | Germany 1- (1961-) | Annual | English/ French/ German | Articles, notes and reviews on cartography. (18 (1978)). |
| KARTOGRAPHISCHE NACHRICHTEN (Deutsche Gesellschaft für Kartographie, Schweizerische Gesellschaft für Kartographie; Osterreichische Kartographische Kommission in der Osterreichischen Geographischen Gesellschaft). | Germany 1-(1951-) | 6 per annum | German | Articles on cartography. News. Discussion. Reviews. (v 29 no 5 (O 1979)). |
| CARTACTUAL: MAP SERVICE BI-MONTHLY. | Hungary 1- (1965-) | 6 per annum | English/ French/ German / | Special maps on timely topics such as new countries, new administrative divisions, dis- |

| | | | | |
|---|---|---|---|---|
| | | | Hungarian | puted areas, pipelines, canals, highways, railroads, air routes, metro lines, bridges, ports, city populations or boundaries, reservoirs, gas fields, power plants, electric transmission lines, reclaimed areas, national parks and reserve areas, and place names. For supplement see Cartinform. (no 79; v 15 (1979) no 5). |
| CARTINFORM: supplement to Cartactual. Publishers' self-reviews (Hungarian National Office of Lands and Mapping. Division of Cartography). | Hungary 1- (Cartactual 27-) (1971-) | 6 per annum | English/ French/ German | Information supplied by publishers on national atlases, regional atlases, other atlases, maps, and books, arranged under each category by country (52 (1979)). |
| UNIVERSO Rivista bimestrale di divulgazione geografica (Instituto Geografico Militare). | Italy 1- (1920-) | 6 per annum | Italian | Beautifully illustrated articles. Cartography. Reports on congresses and meetings. Notes. Book reviews. (v 59 no 5 (S-O 1979)). |
| GEOGRAPHICAL SURVEY INSTITUTE. BULLETIN. | Japan 1- (1948-) | Irregular | English | Mainly surveying, mapping, and aerial photographs but also contains studies of the physical and human geography of Japan. (33 part 2 (Mr 1979)). |

| Title | Country | Frequency | Language | Brief Notes |
|---|---|---|---|---|
| REVISTA CARTOGRAFICA (Instituto Panoamericano de Geografia e Historia Comision de Cartografia). | Mexico 1-23 (1952-73) | Semiannual | Spanish/ English | Technical articles on cartography, geodesy, aerial photos, and remote sensing in the Americas (no 33 (Jn 1978)). |
| CARTOGRAPHIC JOURNAL (British Cartographic Society). | U.K. 1- (1964-) | 2 per annum | English | Articles on cartography. Society record. Recent literature. Recent maps and atlases.Reviews. (v 16 no 1 (Jn 1979)). |
| GEO ABSTRACTS. G. REMOTE SENSING, PHOTOGRAMMETRY, AND CARTOGRAPHY. | U.K. 1-(1974-) | 6 per annum | English | Abstracts of the literature on remote sensing, photogrammetry, and cartography under the following headings: remote sensing general; theory of data acquisition; platforms and instrumentation of data acquisition; data collection platforms; instrumentation for interpretation; data processing; interpretation: general, geology and landforms, soils, snow and ice, water, atmosphere, vegetation and ecology, agriculture and land use, cultural; survey and measurement; photogrammetry and orthophotography; automated mapping: digitising, display, and general; map transformations (projections, |

| | | | | |
|---|---|---|---|---|
| | | | | etc.); map perception; atlases; topographic maps, photo maps; thematic maps; cartography: historical, general. About 2400 abstracts a year. (1979 no 6). |
| IMAGO MUNDI | U.K. 1- (1935-) | Irregular | English | International serial on the history of cartography. Articles. Papers at conferences. Reports and reviews. Obituaries. Bibliography. (v 31, s 2 v 4 (1979)). |
| AMERICAN CARTOGRAPHER (American Congress on Surveying and Mapping). | U.S.A. 1- (1974-) | 2 per annum | English | Articles on all aspects of cartography. Reviews. Recent literature. Cartographic news. (v 6 no 2 (O 1979)). |
| REMOTE SENSING OF ENVIRONMENT An interdisciplinary journal. | U.S.A. 1- (1969-) | 4 per annum | English | Articles. Short communications, book reviews on technology and application of remote sensing to analysis of landscape and environment. (v 8 no 4 (D 1979)). |

| Title | Country | Frequency | Language | Brief Notes |
|---|---|---|---|---|
| REMOTE SENSING QUARTERLY (Association of American Geographers. Remote Sensing Committee; National Council for Geographic Education. Remote Sensing Committee; University of Nebraska at Omaha. Remote Sensing Laboratory and Department of Geography ). | U.S.A. 1- (1979-) | Quarterly | English | Short articles, notes, and news on techniques, applications, and teaching of remote sensing for geographers. Supersedes RSEMS:Remote sensing of the Electro-magnetic Spectrum v 1-5 (1974-1978).(v no 4 (O 1979). |
| SPECIAL LIBRARIES ASSOCIATION. GEOGRAPHY AND MAP DIVISION. BULLETIN. | U.S.A. 1- (1947-) | 4 per annum | English | Articles on maps, map bibliographies, and map libraries. News. Reference aids. New atlases, maps, and books. Reviews. (no 118 (D 1979)). |
| WORLD CARTOGRAPHY (United Nations. Department of Economic and Social Affairs). | U.S.A. 1- (1951-) | Irregular | English | Articles on all aspects of cartography or status of mapping especially on a world wide basis, or for major world regions or problems. Reports- (v 15 (1979)). |

Source: Harris, C.D., Annotated World List of Selected Current Geographical Serials (1980).

Chapter Eleven

# AERIAL PHOTOGRAPHS AND SATELLITE INFORMATION

J.A. Allan

## A Brief History of Remote Sensing

Remote sensing is concerned with the acquisition of data at a distance. The process of acquisition and subsequent use of data requires first that a sensor be directed at some assemblage of physical and human phenomena, secondly that there be a system for recording an image, thirdly that this image can be stored, and fourthly that it can be reproduced. In addition the sensor must be placed in some vehicle or platform which can be navigated appropriately. The most common form of remote sensing is the photographic process carried out from aircraft. Here the camera is the sensor, the aircraft the platform and the film negative the record from which photographic prints can be reproduced as required. In 1972 the range of remote sensing techniques generally available was extended to include a variety of sensing systems carried on satellite platforms. It must be emphasised that despite the impressive specifications of the familiar photographic and the newer sensing systems one of the most versatile and uniquely effective remote sensing systems is the incredibly co-ordinated eye and brain of man himself. No man-made sensor linked to the most advanced computer system has yet simulated the very special capacity which the human eye and brain have for recognising patterns and relating such patterns to the spatial arrangement of natural or artificial phenomena. We shall see later that the specialists who concern themselves with mapping the disposition of features of physical and human geography, the geologist, the geomorphologist, the soil scientist, the vegetation scientist, the forester, and those concerned with rural and urban development each has his own use for imagery. These uses are generally overlapping, or integrated, in that the arrangement of geomorphological, soils and vegetation phenomena are interdependent, but each specialist brings his own expertise to the image and makes maps with different legends.

Remote sensing began with aerial photography. The virtues of aerial surveillance were recognised as soon as the balloon and the camera could be brought together in the American Civil War, and other balloon experiments were conducted by Laussedat in France during the 1850s. Both the World Wars gave aerial photography a boost and gradually there emerged the science of photogrammetry, that is the making of maps based on the precise restitution of the relative position and orientation of the camera at the moments of exposure in the aircraft. The creation of an apparently three dimensional picture by viewing simultaneously two photographs of the same object, exposed at slightly different positions, was familiar in the late nineteenth century, and the principles of <u>stereoscopy</u> were incorporated into photogrammetric instruments so that not only the relative position of terrain features could be observed but much more important their position could be accurately determined, and their height calculated. The degree of precision of maps based on photogrammetric techniques as with all surveying depended on the quality of planimetric and height control available for orientating the three dimensional model created in the viewing devices of the instruments.

The main use of aerial photography has always been, and remains, the mapping of topography, namely the relief features together with the most helpful man-made artifacts, such as main settlements and communications, helpful to a map user for navigation or decision making. Photogrammetric techniques have substantially replaced other methods of detailed survey except in places inaccessible to the aerial view or in circumstances where especially large scale engineering plans are required.

The other important application of aerial photography, that of thematic mapping, increased dramatically during and after World War II. The military uses of aerial photographs created a strong demand for imagery and specialist interpreters. At the end of hostilities the well tried and proven mapping potential of aerial photography was deployed over huge previously unmapped regions of Canada, Australia, Africa and the Middle East. In the 1940s and 1950s aerial photography and map products could be produced relatively inexpensively. By 1970, however, the cost of conventional aerial imagery had become so expensive that its use for resources studies was not justified. Only those studies related to proposed developments with high anticipated economic returns, such as mining or intensive irrigated farming, justified the costs of special flights for the acquisition of photography.

Remote sensing grew out of aerial photography and despite the difficulties of cost referred to in the previous paragraph the field of remote sensing expanded rapidly, especially since 1960 when the term was coined by Evelyn Pruitt. The launch of satellites in the 1960s for military purposes proved the potential of such platforms, and the

Skylab (1973) mission made high quality photographic products available to environmental scientists. The major boost to remote sensing from space for non-military users was the launch of the Landsat programme, originally Earth Resources Technology Satellite (ERTS) in 1972. The successful launch of Landsat 1 was followed in 1975 and 1978 by Landsats 2 and 3. These satellites carried multi-spectral scanning devices on board which rendered reflectance from the terrain digitally, enabling computer manipulation of spatial records to maximise the interpretative and classification potential of the imagery. The Landsat technology was created by the United States; the Soviet Union was the only other nation with the technical competence to launch satellites for remote sensing and high quality photographic imagery has been acquired on behalf of Soviet and Eastern Bloc scientists.

## The Place of Remote Sensing in Geographical Studies

The techniques of photo-interpretation have been familiar to geographers for many decades and they have had a significant place in most geography courses since 1945. As already indicated imagery of all types taken from airborne and spaceborne platforms can be used for two purposes. It can be used to make precise maps showing selected features of relief and topography, and secondly it can be used by specialists in environmental and urban studies to interpret phenomena of importance to them or their clients. The first activity involves careful measurement, is essentially mathematical, is chiefly related to the science of surveying and has a long tradition. Photograph and image interpretation is a relatively new technique, which has been given prominence occasionally, and especially during periods of war, for military purposes. This connection with military applications has raised lasting and unfortunate impediments to the use of aerial imagery for the scientific survey of environmental features such as geology, geomorphology, soils, vegetation, forest resources, cropping, land use as well as settlement geography and urban studies.

Aerial imagery provides the map-like perspective which many types of geographical study require. Images recorded by sensors in aerial and space platforms record a wide range of tangible indicators of ecological and socio-economic circumstances at the surface of the earth. The aerial image is frequently more useful than other spatial records such as the topographic map and for many earth and social scientists requiring a spatial display of selected thematic information the aerial image provides a record unobtainable by any other means at similar cost. Accurately located topographic information, normally absent from remotely sensed imagery, is essential for orientation, however, and is a necessary supplement to thematic information. Such spatial control of imagery is necessary for the accurate derivation of areal estimates which are not possible using unrectified air photographs or orthophotos because of their tilt and height-derived distortions. The distant perspective of space-borne as opposed to aerial sensors provides imagery with negligible spatial distortion and such imagery can be used for estimating and confirming areas provided the resolution of the satellite imagery permits the identification of the phenomena of thematic interest.

Topographic maps provide spatial records of a small number of durable and generally man-made features and the more permanent features of landform represented by contours. Specialists in many environmental and social sciences are interested in phenomena which are ephemeral, with seasonal and annual variations being factors of great importance in explaining the decision making of individual farmers and government officials and the resulting spatial arrangement of human and especially agricultural activity. Aerial and satellite imagery show the type and extent of land cover, whether this is rock, bare soil, natural or semi-natural vegetation or crops. Those disciplines concerned with land cover are therefore in a privileged position in that sensors record the reflectance from crops and vegetation, rather than from the soil and other environmental features beneath. The cover in which they are interested is directly sensed; they are not looking at indicators. Geological and soil features (which may also be of great significance to the geographer) are often obscured by vegetation and crops and specialists have to become proficient in using the disposition of vegetation and land use as indicators of geology, landforms and soils.

Air and space-borne sensors flown at a timely season are not only appropriate for recording ephemeral data, they are especially useful in those parts of the world where the inventory and monitoring of crops and vegetation, and geological and soils mapping are currently inadequate for effective national resource management. Many developing countries lack reliable and up-to-date topographic and resources mapping and it is in those regions that remote sensing from space has a special role.

The use of remotely sensed imagery systems is not without its constraints, and these can be divided into environmental, economic and political problems. The major environmental problem is cloud cover. The familiar photographic and multi-spectral sensors are not all-weather systems. Only micro-wave systems such as radar penetrate cloud, and such imagery has, or will have, a special place in temperate and tropical latitudes where the land surface is obscured for a high proportion of the year or a season. The United Kingdom is particularly troubled by cloud cover and it has proved to be extremely difficult to obtain completely cloud-free cover of the islands for one season never mind for one sequential set of over flights of the satellite. Equatorial countries such as Panama and Indonesia have areas which have never been sensed by Landsat sensors, while countries with massive inventory problems such as Brazil and Nigeria have turned to all-weather aerial radar sensing for comprehensive national surveys of renewable and non-renewable resources. The regions of the world which are most frequently cloud-free are the low-rainfall arid and semi-arid regions of Africa, the Middle East, Asia and North America. (See also Parry and Stanley 1978, and Sabins 1978 pp 177-232).

The second constraint on the use of remotely sensed imagery is the cost of deploying such systems. Photographic survey from aircraft had become a very expensive system by 1980 and so it could only be adopted for studies of tracts with a high economic potential. Mining and oil exploration can readily cover such costs and these have been and remain the main customers for aerial and space imagery. Only agricultural developments which will yield high returns,such as irrigated farming, could support a survey overhead, and the huge tracts of the semi-arid world utilised at very low levels of intensity can only be surveyed by systems which cost very little per square kilometre. The Landsat system is therefore ideal in these semi-arid areas, used generally for grazing, as the system covers large areas very rapidly and at low cost.

The third problem facing those wishing to use and provide library space for remotely sensed imagery is the political sensitivity of such materials. Surveillance from aircraft and satellites has always had military connotations. Many governments have been resentful that Landsat imagery of their countries has been so readily available thereby threatening their national security. There certainly are technologies in the hands of the military organisations of the United States and the Soviet Union which acquire imagery which is extremely intrusive in a security sense. Understandably this imagery is not commercially available and when devising the specification for the first resources satellites a spatial resolution was set at the coarse level of 80 metres,and it has only been with the increasingly wider recognition of the usefulness of satellite monitoring that the spatial resolution has been improved to 40 metres in the RBV (return beam vidicon) system on Landsat 3. Similar levels of resolution are planned for subsequent land resource satellite sensors.

## Image Acquisition and Processing

A number of comprehensive texts are available which describe the aerial and satellite imaging systems (American Society of Photogrammetry 1960, Barrett and Curtis 1976, Rudd 1974, Sabins 1978) and the most succinct and elegant description of the Landsat programme appears in a World Bank publication (1975). There is space here for only a brief introduction to the types of image which can be obtained from specialist countries and which may find their way into lib-

Table 5

Summary of remote-sensing systems

| Image system | Wavelength | Property detected | Operating constraints | Special applications |
|---|---|---|---|---|
| UV imagery and photography. | 0.3 to 0.4 m | Reflectance and flourescence from solar radiation. | Daytime, good weather, low altitude to minimize atmospheric scattering. | Detecting oil films on water and flourescent minerals. |
| Normal colour and black-and-white photography. | 0.4 to 0.7 m | Reflectance of visible solar radiation. | Daytime, good weather, any altitude. | Provides high spatial resolution and stereoscopic coverage. |
| IR colour and IR black-and-white photography. | 0.5 to 0.9 m | Reflectance of visible and photographic IR radiation. | Daytime, good weather, any altitude. | See above: also used for mapping vegetation soil moisture, land-water contacts. |
| Landsat Multi-spectral Scanner. | 0.5 to 1.1 m | Reflectance of visible and photographic IR radiation. | Daytime, good weather, 918 km altitude. | Repetitive regional coverage; available in digital form. |
| Thermal IR imagery. | 3 to 14 m | Radiant temperature. | Day or night, good weather, any altitude. | Mapping temperature variations of land, water and ice. |
| Radar imagery. | 1 to 25 m | Surface roughness and topography. | Day or night, most weather conditions; presently limited to aircraft operation. | Mapping fractures and roughness; especially useful in areas with poor weather. |

Source: Sabins, 1978 p.403.

rary collections.

The photographs from aerial surveys are still the most familiar type of product, but the camera lens and photo-sensitive film is not the only medium for recording surface features from a distant perspective. Since the early 1960s when major experiments with multi-spectral scanners (MSS)began, this system of recording adjacent rectangular picture elements (pixels) of relatively coarse resolution has become increasingly common at least from satellite platforms. The success of the Landsat programme has been based on the imagery produced by the multi-spectral scanners on board the three Landsat satellites.

The black and white as well as colour photographic products should be distinguished from the imagery recorded by multi-spectral scanners. The photographic image is subject to the disadvantages of the relatively unstable chemical process by which it is exposed and processed. The multi-spectral image is recorded pixel by pixel with reflective measures for four or more parts of the visible spectrum recorded on magnetic tape in digital form. This provides a much more versatile image than the conventional photograph and readily lends itself to many quantitative and statistical manipulations.

Most imagery is displayed for the user in photographic form. For example thermal imagery, which utilises the longer wave length portions of the electro-magnetic spectrum, and radar images which record reflected signals in the even longer microwave bands usually come to the user as photographic film or prints, sometimes colour enhanced by computer processing. Potential users of all types of imagery and those who expect to provide library facilities for them should be aware that there is an increasing tendency to display imagery on television monitors and to carry out intermediate analysis and classification procedures. Such manipulations might be achieved by scanning hard-copy photographic pictures or negatives with a television camera for display on a monitor. Otherwise imagery recorded in digital form on magnetic tape can be read and displayed, and generally with better effect than the pictures via the television camera, which are inevitably degraded in resolution and information content. In discussing the modes of displaying imagery it is appropriate in passing to note that devices for viewing imagery and especially for enlarging it enhance the usefulness and accessibility of imagery. Devices familiar to libraries, such as microfilm readers, are often versatile enough to enlarge and display images held in film-positive form. Whilst the usefulness of conventional aerial photography is immeasurably enhanced if there is a stereoscope available to view adjacent photographs in three dimensions.

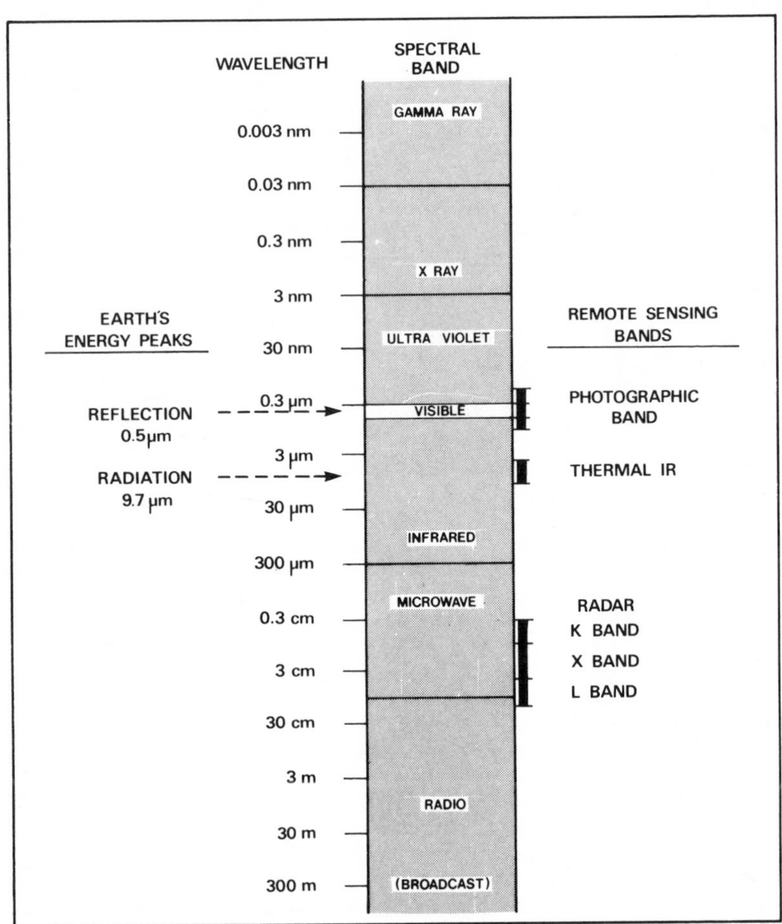


Figure 6. ELECTROMAGNETIC SPECTRUM SHOWING BANDS EMPLOYED IN REMOTE SENSING.

Source: Sabins, 1978, p4.

A high proportion of imagery is recorded in digital form on magnetic tape and in the years ahead this proportion will progressively increase. The tape medium for recording spatial information must take its place amongst the materials handled by libraries serving those concerned with geographical sciences, as such records may provide unique images of areas otherwise without terrain data. Even more important may be the historical information of these spatial data sets. Just as old maps provide important insights into environmental and socio-economic change, so old imagery may provide irreplaceable data concerning terrain modification as well as demographic and economic developments. That a library does not have tape reading facilities should not preclude the acquisition and cataloguing of tapes, especially if there is a professional user community aware of the unique nature of these as yet relatively unfamiliar spatial and temporal records. Digital processing of imagery of all types will, however, be beyond the scope of all but specialist laboratories, at least for some time. Spatial records involve extraordinarily large data sets and although the manipulations of the data may not be complex mathematically, the digital processing of data from for example multi-spectral scanners (MSS) on Landsat satellites, requires either very large general purpose computers, or more recently the dedication of small computers substantially hard-ware orientated, to the task of processing the massive quantities of data gathered by the space-borne MSS scanners. Further developments in software are constantly being effected and the technology was by no means at a final stage at the time of writing in 1980. It is likely that users of imagery will want to deposit in libraries the products of their digital and other image processing, and libraries should consider the place of such material and the cataloguing requirements to ensure its accessibility. That users will come with differing levels of skill in image use should not be a reason for not handling these unfamiliar products. We shall return to this subject in the section concerning image use and interpretation.

## Image Interpretation and Analysis

There is no such scientist as a remote sensing scientist. Successful image analysis requires that engineers and physicists who create the technology relate productively with users, such as environmental and socio-economic scientists. Unfortunately the interface between these specialisations is by no means always elegant or attended with complete mutual understanding, but as the relationships have matured mutual respect has tended to grow, so photographers, engineers and physicists have recognised the futility of producing enhanced imagery which they cannot interpret and,conversely,the geomorphologist, ecologist and specialist in rural development have come to terms with the technology for acquiring and processing imagery. Increasingly there is a tendency to arrange that both types of specialist can contribute to the maximum, by making the systems of image enhancement interactive; that is skilled specialist users can manipulate visual displays by systems devised by physicists and computer scientists, normally via a television monitor to optimise the enhancement of the product for their particular application. Having achieved a suitable enhancement or classification of the image the user then requests a hard-copy of the display, and it is these products which are likely to come into the hands of libraries.

Many books and articles cover the special role of remote sensing in geographical sciences. The 2nd edition of Dickinson's _Maps and Air Photographs_ (Arnold, 1979) lists a large number of these. Stone has contributed a useful historical survey of the ways in which remotely sensed data have been incorporated into geographical methodologies in Estes and Sengal's seminal work _Remote Sensing_ (Hamilton, 1974) and geographers have themselves done their share in publicising the role of black and white photography in environmental studies. They have also commented fully on the special applications of aerial and space imagery to various systematic branches of the discipline: in _geomorphology_ (Verstappen,1974 and 1978), _relief analysis_ (Steiner and Guttermann, 1966), _biogeography_ (Cole, 1965, Cole _et al_, 1974, Cole and Owen-Jones, 1974), _pollution_ (Schneider 1974a and 1974b), _meteorology and climatology_ (Barrett 1970 and 1974) and _land resources_ (Mitchell, 1973).

The aerial image provides simultaneously an overall view of a wide variety of environmental phenomena and as such it is an integrating record. Every type of data from landforms to population and housing quality may be visible. Such an eclectic record is as unselective as many view geographers to be in the range of their interests. The aerial image was, however, for some scientists the inspiration and for others merely the essential tool for the integrated studies which marked much work in land resources surveys in Canada, South Africa and Australia in the 1950s and early 1960s (Christian and Stewart 1968, Stewart 1968). The participation in this work by geographers such as Mabbutt (1968), Verstappen and Steiner, was considerable and they all recognised the integrating character of aerial imagery and especially the landforms visible on them. The importance of the photograph as a basis for surveying and evaluating the natural features of tracts of land has been taken further by a soils specialist (Mitchell 1973b and Mitchell _et al_ 1979) and elaborated by Way (1973). Mitchell shows the air photograph, interpreted by the specialist to be an irreplaceable basis for the simplification of the complex phenomenon which is the natural geographic environment. This simplification is a necessary preliminary to the classification of terrain (or soil, physiography, land or land-type) units prior to appraisal or evaluation. Geographers specialised in soils science have made contributions to the latest thinking on land evaluation in the Framework for Land Evaluation (FAO 1976), and one author of that publication has traced the development of the subject (Young 1973, 1974 and Young and Goldsmith 1979).

A very important feature of raw, as opposed to interpreted products, is the huge data store that they represent. Imagery, normally held in photographic form, but increasingly in digital forms, is a basic source of environmental data to the many specialists experienced in utilising their potential; and each specialist will interpret something different in them. Orthophoto maps are a good example of the type of product which can provide more environmental and socio-economic data than a map of the equivalent area. A user with a high reference level can derive even more from photographs, orthophotos and other imagery than from the cartographic product since so much of the original terrain data are recorded on such imagery. At the same time, the behavioural limitation of much of the potential user community must be recognised and the barriers which are created to ready perception of features of descriptive and explanatory importance on remotely sensed imagery. (Gould and White 1974 pp 119 and 173, Board and Taylor 1977). Many users have spent whole careers assimilating graphic systems used by the map making industry and where possible new products should incorporate familiar forms of the cartographic language. Thus annotation of aerial images with geographic coordinates, some basic topographic information and some place names both enhances the usefulness of the imagery and also makes it a more friendly source to potential users (Allen 1977 p259). Finally it is appropriate to emphasise that raw imagery has a very special role in all sorts of environmental and socio-economic studies, and provided the essential basic information about its location, scale (or flying height and focal length or equivalent of the sensor) and precise date and frequently time of exposure are known, then such records may be more valuable than any other spatial record. Libraries should insist that all these basic reference and orientation details are available before accessing the imagery to their catalogues.

## Indexing Remotely Sensed Imagery

All imagery taken either by professional air survey organisations or by national space survey programmes is carefully referenced. Air photographs are recorded on flight indexes (sometimes known as flight diagrams). These record the flightlines and the position of each photo, or sometimes of each alternative exposure, according to the scale of presentation of the index. On the index will appear some geographical coordinates, the national grid if relevant, the date and time of exposure, the flying height and focal length of the lens used, and also the photo scale although this can easily be deduced from the flying height, the focal length of the lens and the elevation of the terrain. This last is, therefore, an additional item of importance to the potential air photograph user. The most widely disseminated satellite imagery is that from the Landsat programme. These satellites were launched in a near polar orbit and record roughly rectangular images covering 185 x 185 kilometres, that is approximately 34,000 square kilometres. The Landsat platforms have orbited at about 920 kilometres above the surface and circle the earth every 103 minutes, or roughly 14 times per day. Scene data are acquired on the orbital pass from north to south. Repetitive coverage has been achieved by one satellite every eighteen days, and more frequently according to the phasing and operability of the satellites in orbit. These satellites have had the virtue of sun-synchronous orbits, such that they view the earth at the same local time on each pass, which has been roughly 9.42 hours at the equator.

Table 6

Sidelap of Adjacent Landsat Coverage Swaths

| Latitude (degrees) | Image sidelap (per cent) |
|---|---|
| 0 | 14.0 |
| 10 | 15.4 |
| 20 | 19.1 |
| 30 | 25.6 |
| 40 | 34.1 |
| 50 | 44.3 |
| 60 | 57.0 |
| 70 | 70.6 |
| 80 | 85.0 |

Source: Landsat Data User's Handbook, p.5-5. (USGS, 1978).

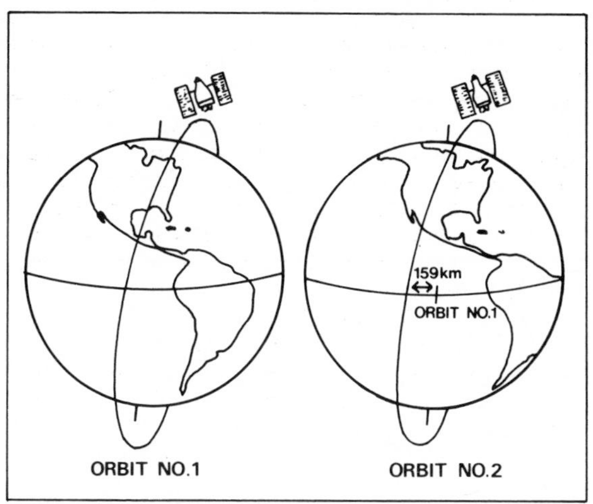


Figure 7. LANDSAT GROUND COVERAGE PATTERN

*The Earth rotates under the satellite causing the ground track of each consecutive orbit to be west of the preceding one.*

Source: LANDSAT DATA USER'S HANDBOOK (1978)

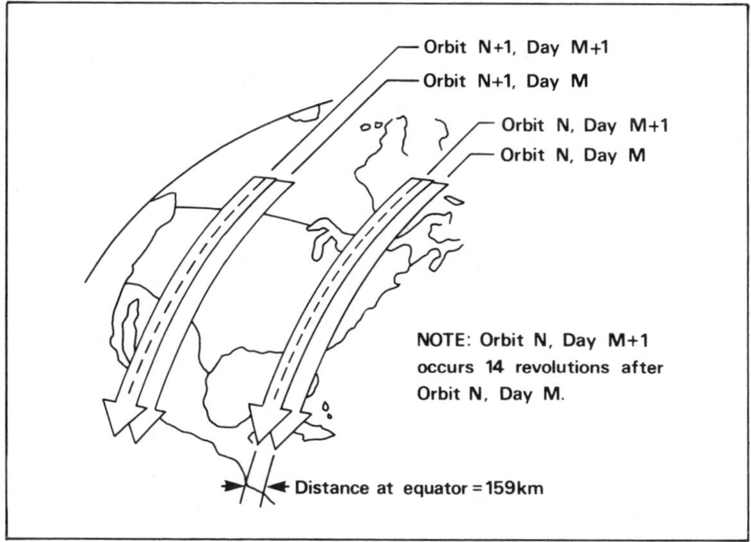


Figure 8. LANDSAT COVERAGE.

*Each succeeding day's orbit overlaps that of the preceding day by an amount that varies depending on the latitude, see table 6.*

Source: LANDSAT DATA USER'S HANDBOOK (1978)

Figure 9. Typical Landsat Ground Trace for One Day
(Only southbound passes shown)

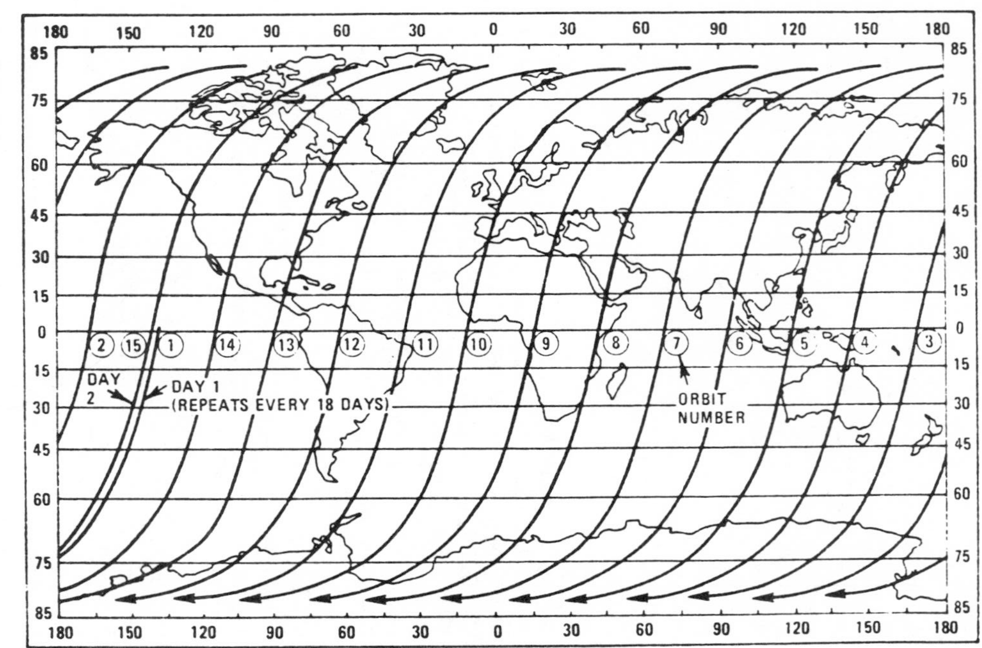


Source: NASA

The digital records have been transmitted directly to receiving stations which have multiplied since the launch in 1972. The first receiving stations were in the United States, but during the 1970s stations were established in Canada, Italy, Iran, Sweden, Brazil, Chile, India, Japan and Australia and more are proposed. During the early years before this network of receiving stations was built up, data had to be recorded on wide band video tape recorders and then transmitted during overflights of the United States within receiving distance of the early receiving stations. Predictably the comprehensiveness of cover has improved greatly since 1976 in Europe with the commencement of operation of the Fucino station close to Rome in Italy. Subsequently the Kiruna station in Sweden has complemented the coverage for northern Europe.

## Sources of Photographs and Images

It is only possible to indicate some of the main organisations concerned with the distribution of aerial and space imagery. Most countries have a number of commercial aerial survey companies and some have national organisations (in addition to a military competence), which can process and distribute air photographs and space imagery. The acquisition of space imagery at the time of writing in 1980 was however limited to the United States and the Soviet Union. However, France and Japan have resources satellites planned, and other nations have used United States satellites for space research and the acquisition of imagery.

Comprehensive collections of aerial photographs are available commercially (Aerofilms Ltd 1974, USDA 1970-71, University of Illinois 1970, National Air Photo Library of Canada 1975). An essential source for those wishing to gain access to the type of aerial images available should acquire the publication Sources of information and materials: maps and aerial photographs (Committee on maps and aerial photographs 1970).

Satellite imagery can be obtained relatively speedily (within three weeks from the United States, and somewhat longer from other stations). Requests for catalogues and computer listings of imagery indicating the quality of the scenes and the degree of cloud cover are turned around even more quickly and those interested in pursuing such procedures are referred to useful articles by Peterson (1975) and Press (1978). The World Bank Atlas (1975) already referred to, provides an invaluable quick guide to the availability of scenes up to 1975 for the developing countries of the world. The Technology Application Center (1975) and Geo Abstracts (1974-1980) provide helpful reviews of literature, and other specialist organisations can provide important reference material (LARS 1974). Also the United States Geological Survey (1972-80) and the United States Department of Agriculture (1975) can supply

Table 7

Satellite remote-sensing systems - 1978 to early 1980s

| System and Date | Sensors | Remarks |
|---|---|---|
| Landsat 3 launched 5 March 1978 | Landsat MSS with addition of a thermal IR band (10.4 to 12.6 m) with a ground resolution cell of 238 by 238 m. | Orbit pattern coverage similar to Landsat 1 and 2. Thermal IR failed 11 July 1978. |
| | Return beam vidicon with a single band (0.5 to 0.75 m) and ground resolution cell of 40 by 40 m. | The return beam vidicon system provides higher resolution images than the MSS. |
| Landsat Follow-on, early 1980s | Same MSS as Landsat 3. | |
| | Thematic mapping scanner with ground resolution cell of 30 by 30 m. In addition to spectral bands in the 0.5 to 1.1 m region there is a band at 0.45 to 0.55 m and a band centered at 1.6 m. | The improved spatial resolution and extended spectral sensitivity of the thematic mapper represent significant technologic advances. |
| Heat Capacity Mapping Mission, launched 26 1978 | Two-band scanning system with 500 by 500 m ground resolution cell. One band in the photographic region (0.5 to 1.1 m) and one in the thermal IR 10.5 to 12.5 m). | Differences in radiant temperature can be determined from day and night thermal IR images. These data can be combined with albedo information from the photographic band to calculate thermal inertia on surface materials. |

| | | |
|---|---|---|
| Seasat A, launched 26 June 1978 | Visible and thermal IR scanning system.<br>Synthetic aperture radar system, = 25 cm.<br>Passive microwave scanning radiometer.<br>Microwave scatterometer. | Systems were designed to record sea state and radiant temperature of the oceans day and night, in all weather conditions. Some experimental imagery of land areas acquired. |
| Space Shuttle | Various imaging systems including cameras, scanners, and radar. | Re-usable shuttle vehicles will carry crews and imaging systems into orbit and return. The shuttle will also serve as a stage for launching small unmanned satellites with imaging systems. |
| Stereosat, planning phase | Return-beam vidicom system will acquire images in a single visible band. | Acquire images with overlap along orbit path for stereo viewing. |

Source: Sabins, 1978 p.403.

worldwide satellite imagery, as can the National Air Photo Library of Canada. Landsat 3 launched in 1978 has on board another type of sensor, the return beam vidicon (RBV) system, which has a 40 metre nominal resolution. Worldwide imagery of selected RBV scenes is also available from the EROS Data Center (United States Geological Survey 1972-80).

In Australia the development of remote sensing has been followed closely. In Australia staff of the CSIRO research centre led the way in using and encouraging the application of aerial survey, and they have involved themselves in satellite image analysis from the first availability of imagery. The Australian ground receiving station became operational in November 1979 and the distribution of images will be coordinated by the Landsat Operation Centre at Canberra. The work of New Zealand scientists can be determined from a review of their Landsat Investigation Programme (DSIR 1979).

Meteorological data is generally acquired at a lower resolution than other resources data and so receiving stations which can handle the lower data rate can almost be assembled by the amateur. A number of university departments in the United Kingdom have such equipment (Dundee, Imperial College London and UCL London) as well as the Royal Aircraft Establishment at Farnborough. Other centres have similar laboratories receiving and distributing data. (National Point of Contact, UK 1980).

The demise of Seasat a few weeks after launch in late 1978 has impaired the research into the use of radar imagery from satellites. A substantial amount of data was received, however, and this can be obtained in the United Kingdom from RAE, Farnborough (1980).

The technology of remote sensing is still at a stage of rapid development. In the field of aerial photography the major development is likely to be in the use of aircraft flying at altitudes up to 20,000 metres. Satellites will also carry high precision cameras for topographic surveys. Meanwhile the planning of the 1980s space platforms is well advanced, and table 7 indicates some of the major programmes.

## REFERENCES


Aerofilms Ltd (1974) Catalogue of stereopairs, Landsat and other aerial imagery also available. Aerofilms, Boreham Wood, England.

Allan, J.A. (1977) 'Map and orthophoto users and their perceptions', Photogrammetric Record, Vol. 9, No.50, pp 257-263.

American Society of Photogrammetry (1960), Manual of photo-interpretation, American Society of Photogrammetry, Falls Church, Virginia.

American Society of Photogrammetry (1976), Manual of Remote Sensing, 2 Volumes, American Society of Photogrammetry, Falls Church, Virginia, 2144 pp.

Barrett, E.C. (1970), 'The estimation of monthly rainfall from satellite data', Monthly Weather Review, Vol. 98, pp 195-205.

Barrett, E.C. (1974), Climatology from satellites, Methuen, London, 418 pp.

Barrett, E.C. and Curtis, L.F. (1976), Introduction to environmental remote sensing, Chapman and Hall, London, 336 pp.

Board, C. and Taylor, R.M. (1977) 'Perception and maps: human factors in map design and interpretation', Transactions of Institute of British Geographers, Vol.2, (New Series) pp. 19-36.

Bundesanstalt für Landeskunde und Raumforschung (1952-74), Geographical interpretation of aerial photographs, BLR Series, Bad Godesberg.

Christian, C.S. and Stewart, G.A. (1968), 'Methodology of integrated surveys', Aerial surveys and integrated studies Proceedings of the Toulouse Conference Unesco, Paris, pp 233-280.

Cole, M.M. (1965), 'The Use of Vegetation in mineral exploration in Australia', Proceedings of the eighth Commonwealth Mining and Metallurgical Congress, Vol. 6 pp 1429-1458.

Cole, M.M. and Owen Jones, E.S. (1974), 'Remote sensing in mineral exploration' in Barrett, E.C. and Curtis, L.F. (editors) Environmental remote sensing, David & Charles, Newton Abbot.

Dickinson, G.C. (1979), Maps and air photographs, Arnold, London.

DSIR (1979), New Zealand Landsat Investigation Programme, Department of Scientific and Industrial Research, Science Information Division, PO Box 9741, Wellington, New Zealand

Estes, J.E. and Senger, L.W. (editors) (1974), Remote sensing: techniques for environmental analysis, Santa Barbara, California, Hamilton Publishing Co. 340 pp.

Geo Abstracts (1974-1980), Section G: Remote sensing and cartography, Geo Abstracts Ltd, University of East Anglia, Norwich, England.

FAO (1976), A framework for land evuluation, Soils Bulletin No. 32, FAO, Rome, 72 pp.
Gould, P. and White, R. (1974), Mental Maps, Pelican, London, 187 pp.
LARS (1974) Bibliography of LARS publications, Laboratory for the Application of Remote Sensing (LARS), Purdue University, Indiana.
Meijerink, A.M. (1969), Photo-interpretation in hydrology:a geomorphological approach, ITC Delft and IPI, Dehra Dun.
Meijerink, A.M. (1974), Photo-hydrological reconnaissance surveys, University of Amsterdam and ITC Enschede.
Mitchell, C.W. (1973a), The application of ERTS 1 Imagery to the FAO Sudan Savanna Project, FAO, Rome.
Mitchell, C.W. (1973b), Terrain evaluation, Longmans, London.
Mitchell, C.W., Webster, R., Beckett, P.H.T. and Clifford, B. (1979), 'An analysis of terrain classification for long range prediction of conditions in deserts' Geographical Journal, Vol. 145, Pt.1, pp.78-85.
National Air Photo Library of Canada (1975), Aerial photograph catalogue (Landsat imagery also available), Ottowa.
National Point of Contact UK (1980), The UK NPOC is located at the Royal Aircraft Establishment, Farnborough. Other NPOCs are located in the countries which are members of the European Space Agency (ESA).
Parry, D.E. and Stanley, D. (1978), 'SLAR for forest survey in Nigeria' in van Genderen, J.L. and Collins, W.G. Remote sensing applications in developing countries. (The Remote Sensing Society, Birmingham).
Parry, D.E. and Trevett, J.W. (1979) 'Mapping Nigeria's vegetation from radar', Geographical Journal, Vol. 145, Pt.2, pp. 265-281.
Peterson, R.A. (1975), 'Techniques for acquiring and employing ERTS 1 photographs for land-use mapping', Professional Geographer, Vol. XXVII, No.2, pp 221-227.
Press, N. (1978), 'Applications, availability and acquisition of earth satellite imagery for geology' in Harvey, A.P. and Diment, J.A., Geoscience information: a state of the art review, Broadoak Press, Heathfield, Sussex, pp. 193-198.
Press (Nigel) Associates (1980), A Landsat world index, compiled and published by Nigel Press Associates, Apex House, 99 High Street, Edenbridge, Kent TN8 5AU. Available on request.
RAE Farnborough (1980) The National Point of Contact at Farnborough distributes unprocessed and processed Landsat and Seasat imagery, Space Department, RAE Farhborough, Hants.
Roscoe, J.H. (1960), 'Photo-interpretation in geography' in Manual of photo-interpretation, American Society of Photogrammetry, Falls Church, Virginia.
Rudd, R.D. (1974), Remote sensing: a better view, Duxbury, North Scituate, Massachusetts, 135 pp.

Sabins, F.F. (1978), Remote sensing: principles and interpretation, Freeman, San Francisco, 426 pp.

Schneider, S. (1974a), Gewasseruberwachung durch fernerkundung - die Mittere Saar, BLR, Bonn-Bad Godesberg.

Schneider, S. (1974b), Luftbild und Luftbildinterpretation, Lehrbuck de Allgemaine Geographie, BD XI, Berlin.

Steiner, D. (1968), Photo-interpretation or national use and conservation of biological resources, Bundesanstalt für Landeskunde und Raumforschung, Bad Godesberg.

Steiner, D. and Gutterman, Th. (1966), Russian data on spectral reflectance of vegetation, soil and rock types, Juris Druck, Zurich.

Stewart, G.A. (editor) (1968), Land evaluation, Papers of a CSIRO symposium in conjunction with UNESCO, August 1968, Macmillan, Canberra.

Stone, K.H. (1974), 'Developing geographical remote sensing' in Estes, J.E. and Senger, L.W. Remote sensing: techniques for environmental analysis, Hamilton Publishing Co, Santa Barbara, pp.1-14.

Technology Applications Center (1975-1980), Quarterly review of literature, Albuquerque, New Mexico. The center also supplies catalogues and satellite imagery, Gemini Apollo, Landsat missions: from TAC, University of New Mexico, Albuquerque, New Mexico, 87131.

United States Geological Survey (1974), Worldwide Landsat imagery and information about coverage can be obtained from EROS Data Center, Sioux Falls, Dakota, 57198, USA.

United States Geological Survey (1979). Data users handbook (revised edition) United States Geological Survey (USGS) from 1200 South Eads St., Arlington, VA 22202.

University of Illinois (1970), Catalogue of University of Illinois air photo repository, University of Illinois, Urbana-Champlain, Illinois.

USDA (1970/1971), An airphoto atlas of the rural United States Agriculture handbooks 372, 384, 406, 409, 419, United States Department of Agriculture, Washington DC.

Verstappen, H.Th. (1968), 'The role of geomorphology in integrating aerial surveys', in Aerial surveys and integrated studies, Proceedings of the Toulouse Conference, Unesco, Paris, p.538.

Verstappen, H.Th. (1970), 'Aerial imagery and regionalisation' Berichte des III International Symposium for Photointerpretation, Dresden.

Verstappen, H.Th. (1974), 'On quantitative image analysis and the study of terrain' ITC Journal Vol. 3, pp 395-413.

Verstappen, H.Th. (1977), Remote sensing in geomorphology, Elsevier Scientific Publishing Co, New York, 214 pp.

Way, D.S. (1973), Terrain analysis: a guide to site selection using aerial photographic interpretation, Dowden, Hutchinson and Ross, Sturdesburg, Pennsylvania, 392 pp.

World Bank (1975), Landsat index atlas of the developing

countries of the world, Washington DC.

Young, A. (1973), 'Rural land development' in Dawson A. and Dournkamp J.C., (eds.) Evaluating the human environment, Arnold, London pp 5-33.

Young, A. (1974), 'The appraisal of land resources' in Hoyle B.S. (ed) Spatial aspects of development, Wiley, London.

Young, A. (1977), 'Soil survey and land evaluation in developing countries: a case study in Malawi', Geographical Journal, Vol 143, Pt 3, pp 407-432.

---

List of European National Points of Contact (NPOC)

Belgium Services du Premier Ministre, 8, rue de la Science, 1040 Bruxelles, Belgium
Tel 02-511 59 85
Twx.24501 PROSCI

Denmark National Technological Library, Center for Documentation, Anker Engelundsvej 1, 2800 Lyngby, Denmark
Tel. 02-88 30 88
Twx 37148 DTBC

Netherlands National Aerospace Laboratory, Anthony Fokkerweg 2, Amsterdam 1017, NL.
Tel. 020-511 31 13
ext 291 Twx 11118

France GDTA, Centre Spatial de Toulouse, 18 Av.Edouard Belin, 31055 Toulouse, France.
Tel. 61-53 11 12
Twx 531081 CNEST A

Germany DFVLR Hauptabteilung Raumflugtrieb, 8031 Oberpfaffenhofen, Post Wessling, Germany.
Tel. 08153-281
Twx 526419 DVL

Ireland National Board for Science and Technology, Shelbourne House, Shelbourne Rd. Dublin 4, Ireland.
Tel. 01-601 797 Twx 30327 NBST

Italy Telespazio, Corso d'Italia 43, 00198 Roma, Italy.
Tel. 06 8497 Twx 610654 TELEDIRO

Spain CONIE, Pintor Rosales 34, Madrid 8, Spain.
Tel. 01-247 98 00
Twx 23495 INVESTE

Sweden Swedish Space Corporation Tritonvaegen 27, 171 54 Solna Sweden
Tel. 08 98 02 00 Twx 17128
MOD PE G

United Kingdom Royal Aircraft Establishment, Farnborough, GU14 6TD, Hants, UK
Tel. 252 24461
Twx 858 134 MOD

Esrin - earthnet via galileo galilei, 00044 frascati, Italy.
(06) 9422401 - Telex 610637 esrini.

Chapter Twelve

# STATISTICAL METHODS AND THE COMPUTER

W.J. Campbell

A recent retrospective and prospective review of quantitative geography has argued that 'whereas the 1960's had often been characterized by early use of technique, the testing of many previously untried tools, and, naturally, the committing of a number of statistical and methodological blunders, the 1970's have seen the emergence of a maturity, a deeper understanding, and an accepted commitment to quantitative work'. (R.J. Bennett and N. Wrigley in chapter one of Quantitative Geography in Britain, (Routledge and Kegan Paul, 1981). A product of this maturity has been the writing of many texts for the newcomer to statistical methods and their applications in geography, in marked contrast to the small handful a decade ago. This guide attempts to indicate the scope and limitations of the primary literature available for students of introductory and further level courses, and is restricted to readily accessible publications available and used in U.K. universities.

These are mainly introductory and more advanced textbooks, written for and by geographers, but attention is also drawn to other sources - for example the CATMOG monograph series (Concepts and Techniques in Modern Geography) edited by members of the Quantitative Methods Study Group of the IBG, and Progress in Geography, now in both human and physical geography series. A full appreciation of many statistical methods requires study of their application, and may also necessitate access to original sources, often in statistics and other disciplines. The references cited provide detailed lists of further reading, source materials, and applications, and the intention here is not to provide such comprehensive cover but rather to indicate useful starting points. Other potential literature sources are likely to be less fruitful; for example the computerized bibliographic data bases available in the U.K. and U.S.A. produced only a few of the references cited here, and there is no recent equivalent of the Bibliography of Statistical Applications in Geography by B. Greer-Wootten (Technical Paper 9, Association of American Geographers, 1972). Two final points of introduction are necessary. This review deals with statistical methods, not the full range of techniques encompassed by mathematical and quantitative geography, nor is it a guide to the application of statistical analysis in many specialist fields of geographical enquiry. Other sources must be consulted for these topics, and a useful starting point might be Quantitative Geography in Britain edited by R.J. Bennett and N. Wrigley (Routledge and Kegan Paul, 1981), a large collection of papers embracing both method and applications over a wide spectrum of quantitative work.

Preliminaries

Students of statistics in geography soon encounter questions which although not concerned directly with specific techniques, affect the way in which such methods are studied and used. Typically these questions revolve around the appropriate role of numerical measurement and analysis and the intrinsic nature of statistical reasoning, and some understanding of these issues is necessary for a proper appreciation of applied statistics. Some of this material is not well represented in introductory texts, since most authors assume that their work will be complemented by lectures and discussion, practical experience and further reading, and so a number of the salient points are taken up here.

The role of statistical analysis is part of the wider philosophical question of the nature of geography as an environmental or social science, a prominent theme in methodological debates for the past two decades. Much of this material is rather abstract and difficult for the newcomer, but readable introductions can be found in Geography: its History and Concepts by A. Holt-Jensen (Harper and Row, 1980), in chapter one of Locational Models by P. Haggett, A.D. Cliff and A. Frey (Volume 1 of Locational Analysis in Human Geography, Arnold, 1977), and in chapters one and two of Themes in Geographic Thought edited by M.E. Harvey and B.P. Holly (Croom Helm, 1981). Introductory statistical texts generally make very limited reference to this issue, but the brief discussions in Quantitative and Statistical Approaches to Geography by J.A. Matthews (Pergamon, 1981) and An Introduction to Quantitative Analysis in Human Geography by M.H. Yeates (McGraw-Hill, 1974) are better than most. Further pursuit of this topic should still make reference to the very influential work Explanation in Geography by D. Harvey (Arnold, 1969).

The nature and process of measurement, and problems of observing and gathering numerical data, also receive little attention in statistical texts. This is surprising in view of the strongly empirical nature of the discipline and the fact that the success of quantitative interpretations rests almost as much on the nature and quality of the data as on the analysis used. Many texts discuss little more than scales of numerical measurement - nominal, ordinal, interval and ratio and sampling procedures, including the sampling of spatially distributed data. Some exceptions to this general rule are worth pursuing - for example, a chapter on compiling geographical information in Patterns in Human Geography by D.M. Smith (Penguin, 1977), a section on the theory and practice of measurement in Quantitative Methods in Geography by P.J. Taylor (Houghton Mifflin, 1977), and chapters on data collection and map description in Locational Methods by P. Haggett, A.D. Cliff and A. Frey (Volume 2 of Locational Analysis in Human Geography, Arnold, 1977). Measurement by example is demonstrated in Quantitative and Statistical Approaches to Geography by J.A. Matthews (Pergamon, 1981), which differs from other texts in being a practical manual and therefore draws attention to the data used in many good, workable problems. The most complete though more advanced discussion of the methodology of measurement and data collection is still part five of Explanation in Geography by D. Harvey (Arnold, 1969), and for the more numerate, section one of Maps and Statistics by P. Lewis (Methuen, 1977) provides an interesting view of maps and measurement.

Fundamental statistical concepts or primitives as distinct from particular methods of analysis are important, for on them rests the rationale, the process of reasoning and argument, supported by classical statistics. Apart from offering methods for more precise and objective description, statistical procedures are designed to permit conclusions (inferences) to be drawn about an investigator's ideas or propositions (hypotheses) in the light of available empirical evidence (data). These inferences are based on statements about likelihood rather than certainty, the principal source of uncertainty usually being the incomplete (sample) nature of available data. The essential working concepts are those of probability and probability distributions, inference and hypothesis testing, and sampling and statistical estimation. All introductory texts deal with these topics, but more stress is laid on them in Statistical Concepts in Geography by J. Silk (Allen and Unwin, 1979) and Inferential Statistics for Geographers by G.B. Norcliffe (Hutchinson, 1977). A good chapter on probability and geographical research inferences can be found in Quantitative Methods in Geography by P.J. Taylor (Houghton Mifflin, 1977). At a rather more advanced level, an interesting approach to statistical concepts for investigating map data can be found in section one of Maps and Statistics by P. Lewis (Methuen, 1977), and chapter fifteen of Explanation in Geography by D. Harvey (Arnold, 1969) contains a good discussion of the methodological aspects of probability and inference.

A common difficulty among those new to statistical analysis is the plethora of alternative techniques and the problem of selecting appropriate methods for the task in hand, and most texts are implicitly aware of the need to help provide organizing typologies. No uniform approach is adopted, but several important distinctions are commonly recognized, for example descriptive and inferential methods; parametric and non-parametric techniques; statistics appropriate to different types of data measurement; univariate, bivariate and multivariate analysis; one-, two- and k-sample situations; spatial as distinct from aspatial or conventional statistics; and the types of problem or proposition handled, such as differences between groups of measurements, association between variables, or the measurement of spatial or temporal independence. Frameworks based on one or several of these distinctions are used in most texts, but almost none offer explicit overall guidance on the selection of appropriate methods. An exception is the brief but useful concluding chapter of Quantitative and Statistical Approaches to Geography by J.A. Matthews (Pergamon, 1981), and Maps and Statistics by P. Lewis (Methuen, 1977) is organized around a table categorizing procedures on the basis of measurement, scale and type of proposition tested. An early effort to understand the basis of these typologies has its reward when new material has to be assimilated, often from scattered sources or from differently organized literature in statistical science or other disciplines.

A postscript to this section concerns the level of mathematical or numerical skill required to work with statistical techniques, an important issue to many in a discipline spanning social science, arts and scientific specialisms. Introductory texts such as Statistics in Geography by D. Ebdon (Blackwell, 1977), Statistical Methods and the Geographer by S. Gregory (4th edn., Longmans, 1978), Quantitative Techniques in Geography by R. Hammond and P. McCullagh (Oxford University Press, 1978), Inferential Statistics for Geographers by G.B. Norcliffe (Hutchinson, 1977) and Statistical Concepts in Geography by J. Silk (Allen and Unwin, 1979) assume no more than the equivalent of Ordinary Level mathematics, and a willingness to apply sound reasoning rather than statistical expertise. Most emphasise the understanding of statistics through applications and problem solving, an approach most clearly exemplified in Quantitative and Statistical Approaches to Geography by J.A. Matthews (Pergamon, 1981), and in texts such as Patterns in Human Geography by D.M. Smith (Penguin, 1977) and An Introduction to 'uantitative Analysis in Human Geography by M.H. Yeates (McGraw-Hill, 1974). These last two also share with Quantitative Methods in Geography by P.J. Taylor (Houghton Mifflin, 1977) a progression from introductory level to topics of intermediate difficulty. Physical geographers and those specializing in areas of strongly quantitative human geography will on the whole be expected to develop a better grasp of numerical and mathematical concepts, and some aptitude in these is required for a full appreciation of intermediate and advanced texts such as Locational Methods by P. Haggett, A.D. Cliff and A. Frey (Arnold, 1977), Multivariate Statistical Analysis in Geography by R.J. Johnston (Longmans,1978), Statistical Analysis in Geography by L.J. King (Prentice-Hall, 1969), Maps and Statistics by P. Lewis (Methuen, 1977), and Computational Methods of Multivariate Analysis in Physical Geography by P.M. Mather (Wiley, 1976). Glossaries of symbols and notation are contained in Inferential Statistics for Geographers by G.B. Norcliffe (Hutchinson, 1977) and Statistical Concepts in Geography by J. Silk (Allen and Unwin, 1979), as well as in Locational Methods by P. Haggett, A.D. Cliff and A. Frey (Arnold, 1977) and Statistical Analysis in Geography by L.J. King (Prentice-Hall, 1969). Almost all texts assume rather than teach basic mathematical skills, although Quantitative Methods in Geography by P.J. Taylor (Houghton Mifflin, 1977) introduces concepts in applied mathematics and probability, and Statistical Analysis in Geography by L.J. King (Prentice-Hall, 1969) includes appendices on geometries, calculus and matrix algebra, the last of these also explained in Computational Methods of Multivariate Analysis in Physical Geography by P.M. Mather (Wiley, 1976). Those using more advanced textbooks and research papers and needing to learn or relearn applied mathematics will find two useful works in Mathematics for Geographers and Planners by A.G. Wilson and M.J. Kirkby (Clarendon Press, 1975) and G.N. Sumner, Mathematics for Physical Geographers (Arnold, 1978), although much of the material covered is not strictly pertinent to statistical analysis.

## Introductory Methods

This section reviews standard techniques which, with the possible exception of multivariate methods, might be considered essential when introducing geographers to the analysis of numerical data. For convenience of presentation a four-fold division is adopted - univariate, bivariate and multivariate methods common in conventional or classical statistics, and spatial statistics which have derived almost exclusively from different sources and represent a unique class of methods for handling spatially distributed data.

Univariate analysis deals with summarizing the properties of single variables (one-dimensional series) and provides inferential methods for testing hypotheses about the nature and behaviour of such data. These concepts are fundamental to all statistical procedures, and in consequence large parts of introductory texts and courses are devoted to univariate methods of analysis.

Central tendency and dispersion measures, representation of variables by different forms of frequency count and distribution, together with the nature of probability distributions, and ways of transforming and comparing observed data to these distributions, are common to all introductory texts. This material is also explained more fully in Analysis of Frequency Distributions by V. Gardiner and G. Gardiner (CATMOG 19, Geo

Abstracts, 1978).

All texts deal with the concepts of hypothesis testing and inference, usually approached via the twin problems of estimation on the basis of incomplete or sample data and comparison of observed data to known probability distributions. Sampling methods and sampling distributions are central to statistical inference, and an extended discussion of Sampling Methods for Geographical Research is contained in CATMOG 17 by C.J. Dixon and B. Leach (Geo Abstracts, 1977).

The application of inferential reasoning to solving empirical problems is demonstrated by a great variety of tests, largely for sample comparisons or for comparing samples to theoretical or hypothesized distributions. The distinction between parametric tests - for which fairly strong assumptions are required about the underlying nature of the data - and non-parametric or distribution-free tests is stressed, as are the types of data measurement to which it is appropriate to apply particular tests. None of the texts has exhaustive coverage, since the total number of tests is very large, but the basic principles are made clear by the selected methods commonly discussed.

Bivariate analysis describes and tests for relationship or dependence between two variables, and provides the key to introducing more complex lines of enquiry. Much the most common method described in some detail is linear correlation and regression. This topic illustrates clearly the need to consult more than one source, for explanations of goodness-of-fit or error measurement, the problems of statistical assumptions, inference and non-linearity, and applications of the method differ significantly from author to author.

Other methods for bivariate analysis are not commonly found in introductory texts, with the exception of temporal dependence or time series which is given limited coverage. Statistical Methods and the Geographer by S. Gregory (4th edn, Longmans, 1978) and Quantitative Techniques in Geography by R. Hammond and P. McCullagh (Oxford University Press, 1978) examine the application of regression to the estimation of trend in elementary time series, An Introduction to Quantitative Analysis in Human Geography by M.H. Yeates (McGraw-Hill, 1974) describes logistic growth curves, Statistical Concepts in Geography by J. Silk (Allen and Unwin, 1979) discusses probabilistic descriptions of temporal trends and Quantitative and Statistical Approaches to Geography by J.A. Matthews (Pergamon, 1981) deals with time dependence via the example of running means.

Non-parametric measures of association provide an alternative approach when, as is often the case, parametric assumptions cannot be met or when ordinal or nominal scale measurement is used. Rank correlation is described in all basic texts, and coefficients of association between nominal scale variables (based on chi-square) are discussed in Quantitative and Statistical Approaches to Geography by J.A. Matthews (Pergamon, 1981), Inferential Statistics for Geographers by G.B. Norcliffe (Hutchinson, 1977), Statistical Concepts in Geography by J. Silk (Allen and Unwin, 1979) and An Introduction to Quantitative Analysis in Human Geography by M.H. Yeates (McGraw-Hill, 1974).

Multivariate analysis provides ways of examining patterns of dependence between three or more variables, but is generally beyond the scope of currently available introductory texts for geographers. While this is arguably correct, it is surprising to find virtually no reference to the problem of interdependence among several variables when discussing limitations on bivariate analysis, for the underlying logical concepts are not difficult. Patterns in Human Geography by D.M. Smith (Penguin, 1977) contains a brief discussion of multiple regression, and both Quantitative Methods in Geography by P.J. Taylor (Houghton Mifflin, 1977) and An Introduction to Quantitative Analysis in Human Geography by M.H. Yeates (McGraw-Hill, 1974) introduce partial correlation and regression and explore a number of technical and methodological issues. Extensive and more advanced treatments are contained in Multivariate Statistical Analysis in Geography by R.J. Johnston (Longmans, 1978) and Statistical Analysis in Geography by L.J. King (Prentice-Hall, 1969), but further reference to these texts, and consideration of other multivariate methods is deferred to the section on advanced topics.

Spatial statistics derive from the special problems of analysing spatially distributed data. Geographers and others in cognate disciplines have been prompted to develop these methods, both from a desire to describe spatial patterns and processes, and from a need to determine and examine situations where the application of conventional statistics may be inappropriate. Elementary spatial statistics are generally used in two ways - to summarize spatial patterns, and to test whether distributions exhibit other than random arrangements, and at the introductory level three interrelated topics are considered - the measurement and collection of spatial data, centrographic measures analogous to central tendency and dispersion measures, and other techniques for analysing point and areal patterns. The first of these, sampling and data collection, has been referred to in an earlier section.

Centrographic statistics for central tendency (for example, mean centre and median centre) and dispersion measures (for example, standard distance and standard deviational ellipse) are described in Statistics in Geography by D. Ebdon (Blackwell, 1977), Quantitative Techniques in Geography by R. Hammond and P. McCullagh (Oxford University Press, 1978), Patterns in Human Geography by D.M. Smith (Penguin, 1977) and An Introduction to Quantitative Analysis in Human Geography by M.H. Yeates (McGraw-Hill, 1974). Orientation or directional statistics are introduced in Inferential Statistics for Geographers by G.B. Norcliffe (Hutchinson, 1977), and a comprehensive treatment of this topic is contained in Directional Statistics by G.L. Gaile and J.E. Burt (CATMOG 25, Geo Abstracts, 1980). Nearest neighbour statistics can also be used as descriptive measures of pattern, and are explained in Statistics in Geography by D. Ebdon (Blackwell, 1977), Quantitative Techniques in Geography by R. Hammond and P. McCullagh (Oxford University Press, 1978), Statistical Concepts in Geography by J. Silk (Allen and Unwin, 1979), Patterns in Human Geography by D.M. Smith (Penguin, 1977) and An Introduction to Quantitative Analysis in Human Geography by M.H. Yeates (McGraw-Hill, 1974). The measurement of shape is not widely covered, but see for example Statistics in Geography by D. Ebdon (Blackwell, 1977), Quantitative Methods in Geography by P.J. Taylor (Houghton Mifflin, 1977) and An Introduction to Quantitative Analysis in Human Geography by M.H. Yeates (McGraw-Hill, 1974).

Considerably more space is devoted to analysing the distribution of point and area patterns to test the proposition of spatial independence or randomness. Nearest neighbour tests and quadrat analysis for point patterns are widely discussed, for example in Statistics in Geography by D. Ebdon (Blackwell, 1977), Quantitative Techniques in Geography by R. Hammond and P. McCullagh (Oxford University Press, 1978), Statistical Concepts in Geography by J. Silk (Allen and Unwin, 1979), Patterns in Human Geography by D.M. Smith (Penguin, 1977) and An Introduction to Quantitative Analysis in Human Geography by M.H. Yeates (McGraw-Hill, 1974), and also at a more advanced level in Statistical Analysis in Geography by L.J. King (Prentice-Hall, 1969) and Quantitative Methods in Geography by P.J. Taylor (Houghton Mifflin, 1977). Spatial autocorrelation or contiguity tests for independence in the distribution of area data are given in Statistics in Geography by D. Ebdon (Blackwell, 1977) and Inferential Statistics for Geographers by G.B. Norcliffe (Hutchinson, 1977), and in some detail in Quantitative Methods in Geography by P.J. Taylor (Houghton Mifflin, 1977) and Statistical Concepts in Geography by J. Silk (Allen and Unwin, 1979), where directional tests of autocorrelation are also provided. The work Maps and Statistics by P. Lewis (Methuen, 1977) is devoted to a detailed and more advanced treatment of point symbol maps and to the analysis of patterns arranged as line symbols, a topic much less commonly discussed - though see chapter six of Patterns in Human Geography by D.M. Smith (Penguin, 1977). A more exhaustive discussion of spatial independence, autocorrelation and point pattern analysis also appears in Locational Methods by P. Haggett, A.D. Cliff and A. Frey (Arnold, 1977). Several ways of exploiting conventional statistics for the analysis of spatial patterns are also reported. For example, methods for areal map comparison or areal association are provided in Inferential Statistics for Geographers by G.B. Norcliffe (Hutchinson, 1977) based on an interesting non-parametric technique, in Quantitative and Statistical Approaches to Geography by J.A. Matthews (Pergamon, 1981) based on correlation measures, and in some detail in Quantitative Methods in Geography by P.J. Taylor (Houghton Mifflin, 1977) in which a variety of approaches are discussed.

Advanced Topics

Extensions of the bivariate linear regression and correlation models frequently provide the starting point for multivariate analysis in advanced courses in geographical methods. Apart from their widespread application, regression models represent the most readily comprehended form of the general linear model, the basis for most multivariate statistical techniques employed by geographers. A review of the model and various extensions entitled 'The general linear model' by J. Silk, together with papers on many aspects of its application, appears in Quantitative Geography in Britain edited by R.J. Bennett and N. Wrigley (Routledge and Kegan Paul, 1981), and a wide-ranging discussion of methodology and applications can be found in 'Statistical methods in physical geography' by D.J. Unwin (Progress in Physical Geography 1, 1977). Some of the technical difficulties encountered are discussed by P.M. Mather and S. Openshaw in 'Multivariate methods and geographical data' (special volume of The Statistician 3/4, 1974).

Good introductions to multiple regression and correlation, and discussion of associated technical problems, are contained in Quantitative Methods in Geography by P.J. Taylor (Houghton Mifflin, 1977), Statistical Analysis in Geography by L.J. King (Prentice-Hall, 1969), and Multivariate Statistical Analysis in Geography by R.J. Johnston (Longmans, 1978), and also in Linear Regression in Geography by R. Ferguson (CATMOG 15, Geo Abstracts, 1978). An advanced treatment of the technical and computational aspects is in Computational Methods of Multivariate Analysis in Physical Geography by P.M. Mather (Wiley, 1976), and extensions dealing with two-way relationships between variables are described by D. Todd in An Introduction to the Use of Simultaneous-Equation Regression Analysis in Geography (CATMOG 21, Geo Abstracts, 1979). Several related multivariate techniques for analysing dependence and interdependence patterns have been applied and systematically described, but these have never found such general acceptance and use. For example, causal analysis and path analysis are suitable for inferring explicit structures of interrelationship among several variables, and have been well known in other disciplines, especially sociology. The methodology is described by D. Harvey in Explanation in Geography (Arnold, 1969), and statistical methods in Multivariate Statistical Analysis in Geography by R.J. Johnston (Longmans, 1978) and Linear Regression in Geography by R. Ferguson (CATMOG 15, Geo Abstracts, 1978). Methods for handling similar logical structures where one or more variables are measured on categorical scales have been more widely used . Analysis of variance techniques for example are discussed in Multivariate Statistical Analysis in Geography by R.J. Johnston (Longmans,1978), and by J. Silk in Analysis of Covariance and Comparison of Regression Lines (CATMOG 20, Geo Abstracts, 1979) and The Analysis of Variance CATMOG 27, Geo Abstracts, 1980). Cause and effect analysis for variables measured on a nominal scale are discussed in Causal Inferences from Dichotomous Variables by N. Davidson (CATMOG 9, Geo Abstracts, 1976), and an alternative approach as yet little used by geographers, via logistic, linear logit and log-linear models, is described by N. Wrigley in Introduction to the Use of Logit Models in Geography (CATMOG 10, Geo Abstracts, 1976) and 'Categorical data analysis' in Quantitative Geography in Britain edited by R.J. Bennett and N. Wrigley (Routledge and Kegan Paul, 1981).

Attempts to impose order and structure on complex systems for which many interacting variables can be observed have been pursued with energy using other multivariate methods, but these are less widely used by geographers than they were formerly. The most popular of these is undoubtedly principal components and factor analysis, techniques capable of identifying groups of interrelated variables and reducing complex structures to simpler forms. Quantitative Methods in Geography by P.J. Taylor (Houghton Mifflin, 1977) and An Introduction to Quantitative Analysis in Human Geography by M.H. Yeates (McGraw-Hill, 1974), and in more detail Statistical Analysis in Geography by L.J. King (Prentice-Hall, 1969) and Multivariate Statistical Analysis in Geography by R.J. Johnston (Longmans,1978), provide introductions to their use in geography. Detailed accounts can be found in Principal Components Analysis by S. Daultrey (CATMOG 8, Geo Abstracts, 1976) and An Introduction to Factor Analysis by J. Goddard and A. Kirby (CATMOG 7, Geo Abstracts, 1976), and a recent review of 'Factor analysis' by P.M. Mather is contained in Quantitative Geography in Britain edited by R.J. Bennett and N. Wrigley (Routledge and Kegan Paul, 1981). Canonical correlation techniques for measuring relationship between a set of criteria variables and a set of predictor variables are described in Multivariate Statistical Analysis in Geography by R.J. Johnston (Longmans,1978), and by D. Clark in Understanding Canonical Correlation Analysis (CATMOG 3, Geo Abstracts, 1975), but to date have been little used. The problem of grouping similar attributes or entities is the theme of classification and cluster analysis techniques for which a selection of methods are described in Statistical Analysis in Geography by L.J. King (Prentice-Hall, 1969), Computational Methods of Multivariate Analysis in Physical Geography by P.M. Mather (Wiley, 1976), and by R.J. Johnston in Classification in Geography (CATMOG 6, Geo Abstracts, 1976) and Multivariate Statistical Analysis in Geography (Longmans,1978). Classification and the regionalization problem are discussed in Locational Methods by

P. Haggett, A.D. Cliff and A. Frey (Arnold, 1977) and by N.A. Spence and P.J. Taylor in 'Quantitative methods in regional taxonomy' (Progress in Geography 2, 1970). The problem of discriminating between proposed classes or groups, and of correctly assigning new observations to an appropriate class, can be tackled by discriminant functions, discussed in Statistical Analysis in Geography by L.J. King (Prentice-Hall, 1969), Multivariate Statistical Analysis in Geography by R.J. Johnston (Longmans,1978) and Computational Methods of Multivariate Analysis in Physical Geography by P.M. Mather (Wiley, 1976).

The study of spatial and temporal series provides another major focus in advanced analysis courses. The early dependence of quantitative geography on classical statistical methods, and on multivariate analysis in particular, has increasingly been complemented by techniques designed to explore and explicitly model the spatial and temporal dimension, and to reduce the reliance on conventional inferential procedures for the analysis of spatial series. The best general text is Locational Methods by P. Haggett, A.D. Cliff and A. Frey (Arnold, 1977). A useful review emphasizing a process oriented approach to spatial analysis, and identifying common strands in diverse methods, has been provided by B.N. Boots and A. Getis in 'Probability model approach to map pattern analysis' (Progress in Human Geography 1, 1977).

Elementary methods for analysing spatial dependence have been reviewed earlier. Comprehensive treatment of the study of point patterns is provided in Maps and Statistics by P. Lewis (Methuen, 1977) and by R.W. Thomas in An Introduction to Quadrat Analysis (CATMOG 12, Geo Abstracts, 1977) and 'Point pattern analysis' in Quantitative Geography in Britain edited by R.J. Bennett and N. Wrigley (Routledge and Kegan Paul, 1981). A good introduction to the analysis of dependence in areal data can be found in Locational Methods by P. Haggett, A.D. Cliff and A. Frey (Arnold, 1977), and a detailed treatment is the theme of Spatial Autocorrelation by A.D. Cliff and J.K. Ord (Pion, 1973). The widely used method for generalizing map surfaces, trend surface analysis, in which the response variable is described as a polynomial regression function of point coordinate locations, is examined in detail in An Introduction to Trend Surface Analysis by D.J. Unwin (CATMOG 5, Geo Abstracts, 1975) and Computational Methods of Multivariate Analysis in Physical Geography by P.M. Mather (Wiley, 1976). Extensions of surface mapping to categorized data are described in Probability Surface Mapping by N. Wrigley (CATMOG 16, Geo Abstracts, 1977). Much early work in the analysis of spatial series derived from methods for time series, but incorporation of the time domain into statistical models of spatial structure and process has only relatively recently become widely discussed in the geographical literature. A good starting point is a collection of review papers on the theme of space-time modelling, covering 'Spatial modelling' by R. Haining, 'Time series analysis' by L. Hepple, 'Spatial time series' by R.J. Bennett and 'Autocorrelation in space and time' by A.D. Cliff and J.K. Ord in Quantitative Geography in Britain edited by R.J. Bennett and N. Wrigley (Routledge and Kegan Paul, 1981). Recent texts have drawn together many formerly scattered ideas and methods, but these generally remain very difficult for all but a small minority of geography students, whether undergraduate or postgraduate - for example Spatial Time Series by R.J. Bennett (Pion, 1979) and Spatial Processes by A.D. Cliff and J.K. Ord (Pion, 1980).

## Computer Applications

Computers have been used by geographers since the earliest developments in quantitative and statistical methodology, and it is commonplace for data manipulation, analysis and graphical tasks to be computer-aided: indeed they are indispensible in research when applying analytical and multivariate methods to anything other than trivial quantities of data. However, as an aid in teaching statistical skills, computing has only begun to make really significant ground in the last five to seven years as access to vastly improved resources has become simpler and suitable computer software has been more readily available. The attractions of computer use - handling realistic data analysis problems, freedom from tedious and error-prone calculation, ability to simulate statistical experiments - have to be balanced against the risk of computation without understanding, and the need to acquire some new skills. In most cases however there are broader educational aims in becoming familiar with computing.

Several textbooks are available for the geographer wishing to learn computer programming and its applications - Computing for Geographers by J. Dawson and D.J. Unwin (David and Charles, 1976), Computer Programming for Spatial Problems by E.B. MacDougall (Arnold, 1976) and Computers in Geography by P.M. Mather (Blackwell, 1976). Examples of programs for statistical calculations are contained in Computing for Geographers and Computers in Geography, but their authors' purpose is not the provision of statistical software to support teaching and research, nor do they suggest that programming skill is the route to acquiring competence in statistics. A few more advanced texts do go significantly further in this direction, not only reviewing statistical methods and their application, but also considering computational aspects and computer implementation. For example, P.M. Mather in Computational Methods of Multivariate Analysis in Physical Geography (Wiley, 1976) publishes many programs, subroutines and functions for the statistical procedures discussed, and Computer and Statistical Techniques for Planners by R.S. Baxter (Methuen, 1976) and Statistics and Data Analysis in Geology by J.C. Davis (Wiley, 1973) also provide routines for many methods relevant to geographical analysis.

This is not the route followed however by the vast majority of students or even many researchers, requiring as it does computer programming competence before any serious statistical analysis. Instead the bulk of statistical computing uses packages devised for the statistically-aware user who requires computation without computing expertise, instructions being given not in a computer programming language but as quasi-English commands that define statistical operations.

Two distinct but overlapping styles of statistical computing - batch and interactive mode - are represented in statistical packages. The first and traditional approach requires a complete set of instructions - defining the data used and statistics requested - to be submitted as a batch computer job, processed, and the printed results returned in due course to the user. The most successful and widely available package of this type is SPSS, described in the SPSS Primer by W.R. Klecka, N.H. Nie and C.H. Hull (McGraw-Hill, 1975), and in the very full if daunting reference manual Statistical Package for the Social Sciences by N.H. Nie, C.H. Hull, J.G. Jenkins, K. Steinbrenner and D.H. Bent (2nd edn., McGraw-Hill, 1975). The package provides extensive data manipulation facilities as well as a wide range of statistical methods, from elementary data description and tabulation to advanced multivariate analysis. An attractive feature of the manual is that it attempts to explain techniques, with examples and references, in terms suitable for the inexpert user. Many other batch packages are in common use in universities, and among the better known are BMDP, a collection of separate statistical programs described in BMDP: Biomedical Computer Programs by W.J. Dixon and M.B. Brown (University of California Press, 1979), and GENSTAT which provides a powerful language both for calling standard procedures and for writing statistical algorithms. Suitable only for more sophisticated users, it is documented in GENSTAT: A General Statistical Package (Numerical Algorithms Group, Oxford, 1980).

An alternative and more recent approach to statistical computing is the interactive or conversational package in which the user conducts a dialogue with the program via a computer terminal, again using an appropriate command language, and receives immediate response and results. This approach is proving popular in teaching, and is closer to the philosophy underlying computer-assisted learning, a field discussed in Computer Assisted Learning in Geography by I.D.H. Shepherd, Z.A. Cooper and D.R.F. Walker (Council for Educational Technology, 1980), which also references a very small number of interactive programs for teaching statistics in geography. In contrast, a large number of packages now available in universities have been developed by statisticians and by other disciplines. For example, MINITAB is an increasingly used package suitable for the user new to both computing and statistics, and exemplifies many of the features of this category of program. An introduction is contained in the MINITAB Student Handbook by T.A. Ryan, B.L. Joiner and B.F. Ryan (Duxbury, 1976), and it is fully described in the MINITAB Reference Manual by T.A. Ryan, B.L. Joiner and B.F. Ryan (Minitab Project, Pennsylvania State University, 1980). The package is easy to use, flexible in its operation and caters for most of the conventional methods required in teaching introductory (and some more advanced) statistics, including the ability to simulate statistical experiments by sampling and generating probability distributions. It does not cater for multivariate methods other than plots, regression analysis, and table generation, though an interesting feature is the inclusion of some methods for the relatively new techniques of exploratory data analysis reviewed for example by N.J. Cox and K. Jones in 'Exploratory Data Analysis' in Quantitative Geography in Britain edited by R.J. Bennett and N. Wrigley (Routledge and Kegan Paul, 1981). The most widely used interactive statistical package is GLIM, particularly strong in the exploration and fitting of linear models including the log-linear and logit models reviewed by N. Wrigley in 'Categorical Data Analysis' in Quantitative Geography in Britain edited by R.J. Bennett and N. Wrigley (Routledge and Kegan Paul, 1981). Suitable for use by the more experienced statistical user, GLIM is described by R.J. Baker and J.A. Nelder in The GLIM System (Numerical Algorithms Group, Oxford, 1978). Other interactive packages in use in universities include FAKAD, developed at Nuffield College Oxford, IDA from the Graduate Business School, University of Chicago, SCSS, a conversational package for use in conjunction with SPSS, and STATPACK developed at Western Michigan University.

The user of statistical computing is of course constrained in the choice of package or computing style by local software and hardware availability in a way not true of published textbook material. The proliferation of small computer systems and increasingly powerful microcomputers however is radically altering the possibilities for computer assistance in teaching and personal research - see for example 'The use of classroom experiments and the computer to illustrate statistical concepts' by J. Silk (Journal of Geography in Higher Education, 3, 1979). However, developments in geography are as yet largely ad hoc and uncoordinated, and there is for instance no well developed package, comparable to the statistical software cited above, for the analysis of spatially distributed data. There are exceptions - for example the texts in computing mentioned at the start of this section, and a certain amount of personally distributed software - but there is still a considerable gulf between current practice in geographical data analysis and the difficulty with which these methods are handled by the novice computer user or student.

## Conclusion

Some final remarks are necessary to help put the literature reviewed here into perspective. The 1970s have seen not only a continuing 'external' debate about the role of statistical and quantitative methods and methodology in the subject as a whole, but also a critical 'internal' reappraisal and development of the now conventional wisdom embodied in many of the works, especially textbooks, cited above. For instance, a long-standing concern about the appropriateness for spatial series of conventional inferential statistics with their assumptions of sample independence and distribution, the logical and inferential problems caused by varying scales of analysis and by the size and shape of areal units, and the relative merits of confirmatory as opposed to exploratory modes of investigation, are but three examples of the technical and methodological issues vigorously debated among practitioners of quantitative geography - see for example the still relevant and readable paper by P. Gould entitled 'Is statistix inferens the geographical name for a wild goose?' (Economic Geography 46, 1970), and 'The modifiable areal unit problem' by S. Openshaw and P.J. Taylor and 'Exploratory data analysis' by N.J. Cox and K. Jones, both in Quantitative Geography in Britain edited by R.J. Bennett and N. Wrigley (Routledge and Kegan Paul, 1981). Likewise, in the context of teaching, there is no concensus view on what type and amount of statistical competence should be taught or on how this can best be achieved - see for example the discussions on mathematical education in geography degrees by R.J. Bennett, R. Haining and A. Wilson, and on objectives and problems in teaching statistical data analysis by N. Cox, E. Anderson and S. Gregory (Journal of Geography in Higher Education 2, 1978). Such continuous critique and reappraisal may appear surprising to many in view of the well established corpus of statistical methods and methodology used throughout the discipline, but these are of course the essential prerequisites and signs of the future vitality of teaching and research in quantitative geography.

Chapter Thirteen

# ARCHIVAL MATERIALS - GOVERNMENTAL AND OTHERWISE*

Ronald E. Grim

"... and on your right is the National Archives Building which was designed by the noted architect John Russell Pope. Inside are the Declaration of Independence, the Constitution, and the Bill of Rights." These are the comments tour guides often make to their tourist-laden buses as they pass the National Archives Building on their tours of Washington D.C. It would not be surprising if some alert tourist were to remark, "You mean they only keep three documents in that large building? What an inefficient use of our tax dollars!" Obviously, there are many other records, most of which are noted for their historical, political, and genealogical research value. However, these archival records are not limited to those research disciplines. There are many records in the National Archives, as well as in state and local archival repositories and historical manuscript collections, that are useful for geographers.

Since archival repositories and historical manuscript collections are usually associated with historical rather than geographical research, this essay will review the secondary sources that identify and evaluate the major archival and historical materials that are potentially useful for geographical research, with particular emphasis on the United States. Those geographers, other than historical geographers, who are familiar with these institutional holdings, are usually aware of only maps and aerial photographs. Although these two types of records are gaining increasing acceptance as valid archival and historical resource materials, the emphasis will be on the more traditional nongraphic sources that are less

---

*Footnote

This bibliographical essay represents a reorganisation and summation of chapters 6-9 in the author's recent annotated bibliography Historical Geography of the United States: a Guide to Information Sources (Gale Research, 1982). Full bibliographical citations and further comments about contents of each publication are included in this guide.

well known to geographers. Specifically, this review will include the recent literature (from approximately the last fifteen years) that describes, catalogues, or analyses the pre-twentieth-century materials in archival repositories and historical manuscript collections that reflect the geographical development of the United States.

Although the term "archives" defines a fairly specific body of records, the term will be used loosely to include both official institutional and organizational archives, as well as manuscript collections. Archives, in a strict sense, refer to the official noncurrent records of a government agency (national, state, or local), a professional organization, a business, or a religious group. Closely related in function is the historical manuscript collection, which is usually subject orientated rather than institution or organization orientated. Materials from a variety of sources, which pertain to a broad subject, time period, or geographic area, are collected. A primary component of these collections is private or personal papers. The historical materials in either an archives or manuscript collection may be in various formats: textual (manuscript correspondence, diaries, or minutes, as well as published documents such as annual reports); graphic (maps, aerial photographs, architectural drawings, still photographs, or motion pictures); aural (sound recordings); or machine readable records. This discussion will not concentrate on the full range of materials; rather the emphasis will be only on manuscript and published textual materials. A few cartographic records that are closely related to textual records will also be mentioned.

The term "archives" also has an historical connotation, but in a strict sense, archives are not limited to the distant past. Although records are normally retired to an archival repository when they are 25, 30, or 50 years old, archival records may be relatively recent, if they are no longer needed for current administrative use. However, the emphasis of this discussion will be on pre-twentieth century materials, drawing heavily on the research experience of historical geographers. In addition, the full range of twentieth-century archival materials (such as manuscript census schedules) may not be available because of privacy restrictions, current agency use, or lack of full archival appraisal.

Archives and historical manuscripts are potentially available for any geographical/political entity, but here comments will be confined to archival material pertaining to the United States. A major emphasis will be on the records of the national government, although records that were created by state, local, or private sources will also be discussed. These comments will not be confined only to the national or Federal period, but will also include pertinent sources from the colonial period.

Geographers and Historical Sources

Archival sources are used frequently by historical geographers, but few have written book length studies or comprehensive surveys of archival sources. The two most comprehensive surveys by geographers pertain to British sources. The first, Geographical Interpretations of Historical Sources: Readings in Historical Geography edited by A.R.H. Baker, J.D. Hamshere, and J. Langton (David and Charles, 1970), consists of twenty articles that were selected for their geographical interpretation and analysis of particular historical source material (1086 Domesday Inquest, taxation accounts, and nineteenth-century industrial and population censuses). In the second work, Historical Sources in Geography (Butterworths, 1979) M.A. Morgan discusses the content, format, and availability of various types of records (poll taxes, inquisitions, land grants, manorial court records, enclosure awards and maps, tithe maps and apportionments, agricultural censuses, probate inventories, parish registers, population censuses, directories, newspapers, and transportation records). He also makes suggestions for the appropriate research and analysis by historical geographers through the use of case studies.

Comprehensive discussions of a variety of archival sources by American geographers have been limited almost exclusively to the colonial period. The most extensive statements have been made by H.R. Merrens in a number of periodical articles. In "Source Materials for the Geography of Colonial America," Professional Geographer 15 (January 1963): 8-11, he describes various types of primary sources (taxable lists, customs statistics, merchant records, newspapers, estate inventories, land surveys, local court minutes, travel accounts, and maps) that are available for reconstructing the historical geography of the original thirteen colonies. The use of narrative and qualitative primary sources (promotional literature, official reports, travel accounts, natural history accounts, and settlers' statements) for reconstructing the physical environment and the settlers'environmental perception are discussed in "The Physical Environment of Early America: Images and Image Makers in Colonial South Carolina," Geographical Review 59 (October 1969): 530-56, and "Praxis and Theory in the Writing of American Historical Geography," Journal of Historical Geography 4 (July 1978): 277-90 (with J.A. Ernst). The use of these records is illustrated in Merrens' reconstruction of one colony's historical geography in Colonial North Carolina in the Eighteenth Century: A Study in Historical Geography (University of North Carolina Press, 1964). Merrens has also published a selection of seventeenth and eighteenth-century documents (travel accounts, estate inventories, government and ecclesiastical reports, merchant and family letters, and a natural historian's account) which illustrate the variety of records that are available for the study of past geographical patterns in The Colonial South Carolina Scene: Contemporary

Views, 1697-1774 (University of South Carolina Press, 1977). An article by another geographer, R.T. Trindell, "The Geographer, the Archives, and American Colonial History," Professional Geographer 20 (March 1968): 98-102, is also useful because it reviews the major guides to archival and manuscript collections.

No comparable comprehensive studies exist which discuss American sources created during the nineteenth and twentieth centuries. There are articles by geographers which analyse one type of source, but they will be discussed later. The most useful guide for historical geographers to nineteenth and twentieth century documents is R.E. Ehrenberg's "Appendix A: Bibliography to Resources on Historical Geography in the National Archives," pp. 315-49, in Pattern and Process: Research in Historical Geography (Howard University Press, 1975). Although this listing pertains to only one institution, it covers such topics as exploration and settlement, population, agriculture, trade and commerce, manufacturing and industry, transportation, urban development, and natural resources.

General Institutional Guides

Because it is not possible to anticipate all the varieties of archival and historical source materials that are of potential use to geographers, the historical researcher should be aware of the various guides and directories that describe archival and manuscript repositories. The most comprehensive directory of archival and historical manuscript repositories in the United States is the National Historical Publications and Records Commission's Directory of Archives and Manuscript Repositories in the United States (National Archives and Records Service, 1978), which supersedes A Guide to Archives and Manuscripts in the United States by P.M. Hamer (Yale University Press, 1961). The holdings of over 2,600 repositories (Federal agencies, state and local archives and historical societies, museums, and corporate, organizational, and university archives) are described in the revised edition. Besides address, hours, and user services, each entry includes a summary statement of holdings and bibliographical references to guides and finding aids. Also included are cross-references to pertinent entries in National Union Catalog of Manuscript Collections (Library of Congress, 16 vols., 1959-1978). By 1978, this ongoing publication included descriptions of approximately 42,200 individual manuscript collections in 1,076 repositories throughout the United States. Since the entries are submitted by the individual repositories, they are neither uniform nor comprehensive. Indices are cumulated periodically and provide references in one alphabetical listing to personal names, place names, subjects, and named historical periods. Before beginning research in any of these institutions the novice should consult Research in Archives: The Use of Unpublished Primary Sources by P.C. Brooks (University of Chicago Press, 1969), which explains the use of archival and unpublished primary sources. Brooks not only describes the variety of historical repositories, he also discusses finding aids, footnotes, copying, access restrictions, and most importantly, the relationship between the researcher and the archivist.

These basic guides can be supplemented by several other publications. The American Association for State and Local History's Directory: Historical Societies and Agencies in the United States and Canada, edited by D. McDonald (11th edn., 1978), does not describe institutional holdings, but it is useful in listing the address, staff, and major programmes of historical societies and related agencies in the United States and Canada. The revised edition of Harvard Guide to American History, edited by F.B. Freidel (Harvard University Press, 1974) is primarily a bibliography of secondary literature on American history, but it also includes several chapters on printed public documents (Federal, state, and local), unpublished primary sources (National Archives, state and local archives, manuscript collections, and foreign archives), microfilmed material, newspapers, and travel literature. Similarly, History of the United States: A Guide to Information Sources, by E. Cassara (Gale Research Co., 1977), is primarily a guide to secondary literature, but it also has a section devoted to archives and manuscript libraries. A good source for news of recent accessions in major archival repositories is the American Archivist, the quarterly journal of the Society of American Archivists. Although this periodical is devoted primarily to archival administration, it does include occasional topical articles discussing various types of records such as urban, environmental, or immigration-related records. The Society of American Archivists has also published several useful directories including Directory of College and University Archives in the United States and Canada (1980) and A Directory of State Archives in the United States (1980).

State, local, and religious archives, which are widely dispersed throughout the country, were inventoried by the Work Projects Administration's Historical Record Survey during the 1930's. The publications that resulted from these surveys are listed in Bibliography of Research Projects Reports: Checklist of Historical Records Survey Publications, by S.B. Child and D.P. Holmes (Work Projects Administration, 1943). Most of these publications, which had limited distribution, are available in the records of the Work Projects Administration (Record Group 69) in the Legislative and Natural Resources Branch in the National Archives. Related materials in other repositories are described in The WPA Historical Records Survey: A Guide to the Unpublished Inventories, Indexes, and Transcripts (Society of American Archivists, 1980). The Historical Records Survey is also described by

L. Rapport, "Dumped from a Wharf into Casco Bay: The Historical Records Survey Revisited," American Archivist 37 (April 1974): 201-10, and D.L. Smiley, "A Slice of Life in Depression America: The Records of the Historical Records Survey," Prologue 3 (Winter 1971): 153-59.

There are also repositories outside the United States that contain historical source materials relating to the historical and geographical development of the United States. The largest number of these documents are in British repositories and pertain to the colonial period. The most recent guide to British sources is A Guide to Manuscripts Relating to America in Great Britain and Ireland, by J.W. Raimo (Meckler Books for the British Association for American Studies, 1979), which provides brief descriptions of all manuscript collections in Great Britain and Ireland that pertain to the history of the American colonies and the United States. An earlier edition, which was prepared by B.R. Crick and M. Alman, was published by Oxford University Press in 1961. These two editions are supplemented by several guides published by the Carnegie Institution of Washington, D.C. in the early twentieth century: A Guide to the Materials for American History to 1783, in the Public Record Office of Great Britain, by C.M. Andrews (2 vols., 1912; 1914); Guide to the Manuscript Materials for the History of the United States to 1783, in the British Museum, in Minor London Archives, and the Libraries of Oxford and Cambridge, by C.M. Andrews and F.G. Davenport (1908); and Guide to the Materials in London Archives for the History of the United States since 1783, by C.O. Paullin and F.L. Paxson (1914). Some of the original thirteen states have made an attempt to copy those records in foreign repositories that document their development during the colonial period. One of the more ambitious projects on the state level is the Virginia Colonial Records Project. The survey reports and accompanying microfilms are available at the University of Virginia's Alderman Library, Virginia State Library, and Colonial Williamsburg's Research Library. The records in the British Public Record Office which pertain to Virginia are summarized in The British Public Record Office: History, Description, Record Groups, Finding Aids, and Materials for American History with Special Reference to Virginia (Virginia State Library, 1960).

Many significant manuscript collections are available through letterpress publications or microfilm programmes. Some particularly successful publication programmes are the projects sponsored by the National Historical Publications and Records Commission. Over two hundred letterpress and microfilm publications are described in the National Historical Publications and Records Commission Publications Catalog (National Archives and Records Service, 1976). Although most of these projects pertain to the papers of prominent American political figures or corporate bodies, there are several projects that pertain to prominent individuals of geographical interest (Charles C. Fremont, Benjamin Latrobe, Stephen H. Long, Frederick Law Olmsted, Zebulon M. Pike, Henry R. Schoolcraft, and Isaac I. Stevens). An earlier guide to microfilmed records is Guide to Photocopied Historical Materials in the United States and Canada, edited by R.W. Hale (Cornell University Press, 1961). The listings for most states include censuses, government, county, local, church, personal, and business records. Microfilm copies of many of the county or local records are available through the Genealogical Society Library, Salt Lake City, Utah. Their services and holdings are described more fully by L.T. Wimmer and C. Pope, "The Genealogical Society Library of Salt Lake City: A Source of Data for Economic and Social Historians," Historical Methods Newsletter 8 (March 1975): 51-88, and L.R. Gerlach and M.L. Nichols, "The Mormon Genealogical Society and Research Opportunities in "Early American History", William and Mary Quarterly, 3rd. series, 32 (October 1975): 625-29.

Guides to Major Repositories

Most archival and manuscript repositories engage in description programmes that result in published inventories of specific collections or their entire holdings. Although many of the more comprehensive institutional finding aids are listed along with the institutional descriptions in Directory of Archives and Manuscript Repositories in the United States (1978), special reference will be made to several of the major repositories. The holdings of these repositories are national in scope, and a discussion of the pertinent finding aids will illustrate the type of guides that are available for other institutions.

The National Archives and Records Service, which is the official repository for the noncurrent records of the various agencies of the United States government, consists of the National Archives Building in downtown Washington, D.C., fifteen regional archives and records centres, and seven presidential libraries. The most comprehensive listing of its holdings is the Guide to the National Archives of the United States (National Archives and Records Service, 1974). This publication describes records from over 400 agencies that were accessioned as of June 30, 1970. For each record group (the designation for the records of an individual agency, bureau, or commission), there is a brief administrative history; a summary description of the textual, cartographic, and audiovisual records; and a listing of the published finding aids and microfilm publications. Besides the numerous preliminary inventories, special lists, and reference information papers that describe various segments of these records, the National Archives has sponsored a number of conferences that were designed to acquaint specific academic groups with the research potential of the National Archives' holdings.

One of these conferences was for historical geographers. The proceedings of this conference, which was held in November 1971, were published as _Pattern and Process: Research in Historical Geography_, edited by R.E. Ehrenberg (Howard University Press, 1975). Ehrenberg's appendix, in which he listed the published finding aids that would be of interest to historical geographers, is included in this volume. The proceedings of other conferences that may be useful to historical geographers include: _United States Polar Exploration_, edited by H.R. Friis and S.G. Bale, Jr.(Ohio University Press, 1970); _The National Archives and Statistical Research_, edited by M.H. Fishbein (Ohio University Press, 1973); _The American Territorial System_, edited by J.P. Bloom (Ohio University Press, 1973); _The National Archives and Urban Research_, edited by J. Finster (Ohio University Press, 1974); _Indian-White Relations: A Persistent Paradox_, edited by J.F. Smith and R.M. Kvasnicka (Howard University Press, 1976); _Farmers, Bureaucrats, and Middlemen: Historical Perspectives on American Agriculture_, edited by T.H. Peterson (Howard University Press, 1981); and _Afro-American History: Sources for Research_, edited by R.L. Clarke (Howard University Press, 1981). While the major emphasis of the guide and the conference proceedings is on the records in the National Archives Building, general descriptions of the records in the regional archives are provided by N.D. Tutorow and A.R. Abel, "Western and Territorial Research Opportunities in Trans-Mississippi Federal Records Centers," _Pacific Historical Review_ 40 (February 1971): 501-18, and A.M. Campbell, "Reaping the Records: Research Opportunities in Regional Archives Branches," _Agricultural History_ 49 (January 1975): 95-99. The National Archives also publishes a quarterly journal, _Prologue: The Journal of the National Archives_, which includes articles based on research in the National Archives, genealogical notes, and news of accessions, declassifications, and publications. Several of the articles have highlighted cartographic records; one particularly useful article, which uses both textual and cartographic records to recreate the local history of one Wisconsin township, was prepared by J.F. Smith, "The Use of Federal Records in Writing Local History: A Case Study," _Prologue_ 1 (Spring 1969): 29-51.

The Library of Congress, which is known to most scholars for its collection of published secondary books and to geographers for the largest map collection in the country, also has an extensive collection of historical manuscripts that are in the custody of the Manuscript Division. This collection, which is national in scope, includes transcripts of documents from British repositories that pertain to the administration of the thirteen colonies, presidential papers that were deposited with the Library of Congress before the establishment of the National Archives, and a wide range of private and business records. Many of the specific collections are included in _National Union Catalog of Manuscript Collections_, while selected manuscript collections are described in Special Collections in the Library of Congress: A Selective Guide, compiled by A. Melville (Library of Congress, 1981). The transcripts from the British repositories are listed in A Guide to Manuscripts Relating to American History in British Depositories Reproduced for the Division of Manuscripts of the Library of Congress, compiled by G.G. Griffin (Library of Congress, 1946). The collections, programmes, acquisitions, and services of the Library are featured in the Quarterly Journal of the Library of Congress, while a summary overview of the Library's entire holdings is provided by the recent publication, Treasures of the Library of Congress, by C.A. Goodrum (H.N. Abrams, 1980).

Although it is impossible to list all major repositories of historical manuscripts, several of the more prominent libraries maintain significant collections of geographic and cartographic interest. The Henry E. Huntington Library in San Marino, California, has over one and a half million sixteenth through twentieth-century manuscripts pertaining to the social and cultural history of the United States. These materials, which concentrate on the Revolutionary War, the Civil War, and Western history, are described in Guide to American Historical Manuscripts in the Huntington Library (Henry E. Huntington Library and Art Gallery, 1979). Manuscript collections in the William L. Clements Library, University of Michigan, Ann Arbor, are listed in Guide to the Manuscript Collections of the William L. Clements Library, edited by J.C. Dann and A.P. Shy (3rd edn, G.K. Hall, 1978). A major component of these collections is British and American material relating to the colonial and Revolutionary War periods and the War of 1812. The Newberry Library, Chicago, holds a variety of material relating to the American Indian, Western exploration and settlement, and the history of cartography. Guides to specific collections include A Check List of Manuscripts in the Edward E. Ayer Collection, compiled by R.L. Butler (Newberry Library, 1937), and A Catalogue of the Everett D. Graff Collection of Western Americana, compiled by C. Storm (University of Chicago Press for Newberry Library, 1968). Manuscript material pertaining to California, other western states, and Mexico are collected by the Manuscript Division, Bancroft Library, University of California, Berkeley. Part of these collections are described in A Guide to the Manuscript Collections of the Bancroft Library, Vol.1, Pacific and Western Manuscripts, edited by D.L. Morgan and G.P. Hammond (University of California Press, 1963). A general guide to the various collections of rare books and manuscripts at Yale University, New Haven, Connecticut, is found in The Beinecke Rare Book and Manuscript Library: A Guide to its Collections (Yale University Library, 1974). Yale's collection of Western Americana is described in more detail in A Catalogue of the Frederick W. and Carrie S. Beinecke Collection of Western Americana, edited by A. Hanna

(Yale University Press, 1965), and A Catalogue of Manuscripts in the Collection of Western Americana Founded by William Robertson Coe, Yale University Library, compiled by M.C. Withington (Yale University Press, 1952). In contrast to these traditional manuscript repositories, the Inter-University Consortium for Political and Social Research in Ann Arbor, Michigan, provides a central repository for machine readable social science data. Its Guide to Resources and Services, 1978-1979, describes the services and holdings: election data (county and state level voting data for major offices from 1789 to present); U.S. census data (county and state level variables from 1790 to 1970); and U.S. Congressional roll call votes (1790 to present).

One type of source that is available in both the National Archives and the Library of Congress, as well as in other major research libraries throughout the country, is Federal government publications. Nineteenth-century documents of particular interest to geographers include Congressional documents, annual reports of the executive departments and bureaus, and reports of exploring expeditions. The basic guide to these publications is Government Publications and their Use, by L.F. Schmeckebier and R.B. Eastin (Brookings Institution, 1969). One particularly important set of documents is the U.S. Congressional Serial Set, which is an ongoing collection of publications compiled under the direction of Congress. This series started with the American State Papers and includes Congressional journals, reports, and manuals and the annual reports of Federal executive agencies. The serial set is available from the Congressional Information Service on microfiche as C.I.S. U.S. Serial Set on Microfiche and is indexed in C.I.S. U.S. Serial Set Index, 1789-1969 (36 vols., 1975-79). More specialized bibliographies of government publications have been compiled. For example, List of Publications of the Bureau of American Ethnology (Smithsonian Institution, Bureau of American Ethnology Bulletin 200, 1971), includes nearly 100 years of publications on the ethnology, linguistics, archaeology, and physical anthropology of the American Indian. Bibliographies pertaining to census publications and government exploration accounts are mentioned in the following sections.

## Types of Archival and Historical Sources

Just as it is impossible to list guides for all repositories, it is also impossible to list all types of records that will be useful for research in historical geography. The nature of a research problem will define the types of records needed, although not always available. The intent of this section is to describe several types of records that have proven useful to historical geographers and to list the appropriate finding aids. These records are classified as population records, property records (real and personal), travel and exploration accounts, and commercial records.

## Population Records

Population records, such as censuses, are useful to geographers who are interested in population distribution and density. Depending on the information collected in a census, these records can also be useful for determining population composition (ethnicity, race, age, sex, family size, or occupation) and population origin (state or foreign country). Agricultural and manufacturing censuses, when available, have also proven useful in geographic studies.

The most familiar and probably the most useful source of population information is the Federal government's decennial population censuses. Beginning in 1790, the U.S. Government enumerated the population for the purpose of determining representation in the Congress. The information gathered in the first census was rather brief (name of head of household and number of free and non-free members in a household), but subsequent decennial censuses have collected additional information. For example, the 1850 census started listing names of family members as well as age, sex, occupation, and birthplace.

The statistical results of the censuses have been reported for the nation as a whole, and for individual states, counties, and/or minor civil divisions. These tabulations, which are useful for mapping broad geographical patterns are available in a variety of publications. These decennial reports, as well as other special publications, are listed in Bureau of the Census Catalog of Publications, 1790-1972 (Bureau of the Census, 1974). This catalogue not only lists 60,000 reports published since 1946, but also incorporates an earlier edition which was entitled, Catalog of United States Census Publications, 1790-1945, by H.J. Dubester (Library of Congress, 1950). Some of the special publications that are useful to historical geographers are retrospective compilations of selected statistics. The most comprehensive from an historical perspective is Historical Statistics of the United States, Colonial Times to 1970 (bicentennial edition, Bureau of the Census, 1975). The first twenty-three chapters cover such topics as population, vital statistics, migration, labour, income and wealth, agriculture, forestry and fisheries, minerals, land sales, manufactures, transportation, communication, and energy, while the final chapter is devoted to colonial and pre-Federal statistics. Another useful publication is A Century of Population Growth from the First Census of the United States to the Twelfth, 1790-1900 (Bureau of the Census, 1909), which is based primarily on a comparison of the 1790 and 1900 census figures. Examples of topical publications dealing with specific population groups are Indian Population in the United States and Alaska, 1910 (Bureau of the Census, 1915), and Negro Population, 1790-1915 (Bureau of the Census, 1918). Also of interest to geographers are the statistical atlases that were prepared to accompany the published reports of the 1870-1920 censuses. The first two of these statistical atlases are reviewed by L.J. Cappon in "The Historical Map in American Atlases," Annals of the Association of American Geographers 69 (December 1979): 622-34. Population statistics from 1790 to 1970 for each state, county, and city (with a 1970 population over 10,000) are also abstracted in Population Abstract of the United States, by J.L. Andriot (Andriot Associates, 1980).

While the published reports and atlases are useful for studying the broad spatial patterns of population, the manuscript census schedules on which the field data was recorded, have proven useful for microstudies of population and social characteristics within areas for which published statistics are not available. Currently, the 1790-1900 census schedules are open to general research. The utility of the 1850-80 census schedules for historical research, and particularly for research in historical geography, is discussed in Federal Census Schedules, 1850-80: Primary Sources for Historical Research, Reference Information Paper No.67, by C.R. Delle Donne (National Archives and Records Service, 1973), while their content and utility for local history are described by R.G. Barrows in "The Manuscript Federal Census: Source for 'New' Local History," Indiana Magazine of History 69 (September 1973): 181-92. A broader publication, which provides a union list of the microfilm and manuscript copies of the population and mortality censuses as well as a discussion of the content of each census schedule, is Federal Population and Mortality Census Schedules, 1790-1890, in the National Archives and the States: Outline of a Lecture on their Availability, Content and Use, Special List No. 24, compiled by W.N. Franklin (National Archives and Records Service, 1971). The content of the 1790-1880 census schedules is also summarized in the genealogical research aid Survey of American Census Schedules, by E.K. Kirkham (Deseret Book Co.,1959). The 1790-1900 census schedules are available on microfilm. The content of the microfilm reels is listed in Federal Population Censuses, 1790-1890: A Catalog of Microfilm Copies of the Schedules (National Archives and Records Service, 1975), and 1900 Federal Population Census: A Catalog of Microfilm Copies of the Schedules (National Archives and Records Service, 1978). The records of the Census Bureau are not confined to the published reports and the manuscript census schedules. Other administrative records in the National Archives are described in Preliminary Inventory of the Records of the Bureau of the Census, Preliminary Inventory No. 161, compiled by K.H. Davidson and C.M. Ashby (National Archives and Records Service, 1964). The census Bureau's cartographic records, which include census enumeration district maps and descriptions, are listed in Preliminary Inventory of the Cartographic Records of the Bureau of the Census, Preliminary Inventory No. 103, compiled by J.B. Rhoads and C.M. Ashby (National Archives and Records Service, 1958).

A number of periodical articles have been written by geographers and historians exploring the accuracy and spatial components of the manuscript census schedules. Articles in the Historical Methods Newsletter include E.K. Muller, "Town Populations in the Early United States Censuses: An Aid to Research," 3, no.2 (March 1970): 2-8 (uses occupational data from the census schedules to estimate population of small towns not recorded in published reports); P.K. Knights, "A Method for Estimating Census Under-Enumeration," 3, no.1 (December 1969): 5-8, and "Accuracy of Age Reporting in the Manuscript Federal Censuses of 1850 and 1860," 4 (June 1971): 79-83; J.P. Radford, "A Note on the Spatial Analysis of Sets of Highly Disaggregated Data," 5 (June 1972): 115-17 (use of fire insurance maps to map data associated with street addresses, such as data obtained from city directories or census schedules). Two articles in Professional Geographer discuss the mapping of census data within the county unit: J.A. Hazel, "Semimicrostudies of Counties from the Manuscripts of the Census of 1860," 17 (July 1965): 15-19 (discusses inherent clues in 1860 schedules for mapping slaveholding patterns within Alabama counties); and M.P. Conzen, "Spatial Data from Nineteenth Century Manuscript Censuses: A Technique for Rural Settlement and Land Use Analysis," 21 (September 1969): 337-43 (uses township landownership plats to map data from population and agricultural censuses). Also useful are three articles in Historical Geography Newsletter: J.E. Westfall, "Estimating Minor Civil Division Boundaries Through the Manuscript Census Schedules: A Methodological Note," 3, no.1 (Spring 1973): 3-6; K.N. Conzen, "Mapping Manuscript Census Data for Nineteenth Century Cities," 4, no.1 (Spring 1974): 1-7 (use of city directories and fire insurance maps in mapping manuscript census data); and J. Sobel, "Population Linkage and the Manuscript Census," 4, no.1 (Spring 1974): 8-12.

Besides the Federal censuses, individual states and territories periodically prepared censuses of their populations. A bibliography of state censuses is found in State Censuses: An Annotated Bibliography of Censuses of Population Taken After the Year 1790 by the States and Territories of the United States, by H.J. Dubester (Library of Congress, 1948).

The manuscript schedules from both Federal and state censuses have been used by geographers to study population migrations and residential patterns in rural and urban situations. For example, T.G. Jordan uses the 1850 Federal census schedules for determining regional and ethnic population origins in Texas in German Seed in Texas Soil: Immigrant Farmers in Nineteenth-Century Texas (University of Texas Press, 1966), and "Population Origins in Texas, 1850," Geographical Review 59 (January 1969): 83-103. Similarly, R.L. Nostrand uses 1850 and 1900 census data in mapping Hispanic population patterns in the southwestern United States in "Mexican Americans Circa

1850," Annals of the Association of American Geographers 65 (September 1975): 378-90, and "The Hispano Homeland in 1900," Annals of the Association of American Geographers 70 (September 1980): 382-96. On the other hand, M.P. Conzen uses an 1895 state census to study geographic mobility within Iowa in "Local Migration Systems in Nineteenth-Century Iowa," Geographical Review 64 (July 1974): 339-61. Manuscript census schedules are also useful for historical and geographical studies of urban social patterns. In particular, geographers have been interested in residential patterns within nineteenth-century cities. Examples include P.A. Groves and E.K. Muller, "The Evolution of Black Residential Areas in Late Nineteenth-Century Cities," Journal of Historical Geography 1 (April, 1975): 169-91 and J.P. Radford, "Race, Residence, and Ideology: Charleston, South Carolina in the Mid-Nineteenth Century," Journal of Historical Geography 2 (October 1976): 329-46.

Population data for the colonial period is not as readily available or as systematic and comprehensive in coverage as the Federal censuses. Population data, whether in the form of censuses (actual enumerations), taxable and militia lists, or general estimates, were generated at various colonial administrative levels and for a variety of purposes. The standard reference for colonial population sources is American Population Before the Federal Census of 1790, by E.B. Greene and V.B. Harrington (Columbia University Press, 1932). One hundred and twenty-four censuses covering twenty-one British colonies provide the basis for The Population of the British Colonies in America Before 1776: A Survey of Census Data, by R.V. Wells (Princeton University Press, 1975). The introductory chapter discusses the sources of these censuses, their content, creation, reliability, and utility, while the remaining chapters are devoted to a demographic analysis of the data. The problems of estimating colonial population figures are addressed by H.A. Whitney in "Estimating Precensus Populations: A Method Suggested and Applied to the Towns of Rhode Island and Plymouth Colonies in 1689," Annals of the Association of American Geographers 55 (March 1965): 179-89. Although American sources are not emphasized, Historical Demography, by T.H. Hollingsworth (Cornell University Press, 1969), is useful for identifying non-census sources of population information. Particular attention is paid to taxation returns, vital registration data, genealogies, wills, military data, and nonwritten sources (pictures and archaeological discoveries).

There are several secondary studies that are geographical in presentation and based extensively on colonial population sources. The footnotes of these publications will serve as guides to the availability and utility of these sources. In his classic study, A Series of Population Maps of the Colonies and the United States, 1625-1790 (rev. edn, American Geographical Society, 1968), H.R. Friis reconstructed ten population dot maps that show the changing population distribution of the

Atlantic Seaboard from 1625 to 1790. Similarly, S.H. Sutherland prepared three population dot maps that show population distribution in the thirteen colonies during the pre-Revolutionary era (1771-1776) in Population Distribution in Colonial America (Columbia University Press, 1936; reprinted, AMS Press, 1966). A synthesis of existing studies provides the basis for tabulating the number of slaves involved in the trade to the Caribbean and North American colonies in The Atlantic Slave Trade: A Census, by P.D. Curtin (University of Wisconsin Press, 1969).

Property Records

The process of surveying and disposing of the public domain in the United States has generated a wealth of records that are useful to historical geographers. Land survey records, which consist of field notes and survey plats (plans of ownership), have a primary legal value for establishing boundaries of a parcel of land, as well as a secondary value for reconstructing vegetation patterns from witness trees and related comments recorded in the notes. The land disposal records, which consist of patents or deeds, land entry papers, and tract books, document the sale and resale of the land and are useful for reconstructing the land ownership process. In identifying land survey and ownership records, a distinction should be made between the public and nonpublic land states. The public land states are those in which the Federal government was the original owner of the land. Because the Federal government was responsible for the initial survey and sale of the land, the pertinent records are in the custody of the Federal government, either with the National Archives or the Bureau of Land Management (successor to the original General Land Office). In the nonpublic land states, which include the original thirteen states, Maine, Vermont, West Virginia, Kentucky, Tennessee, and Texas, the original ownership was vested in the colonial or state government. Consequently, the records pertaining to the original survey and sale of the land are in the custody of a state office, either a state land office or state archives. For both the public and nonpublic land states, the sale or transfer of land subsequent to the initial sale of the land was a local function, usually at the county level. These records have been normally maintained in a county clerk's office and often suffered the ravages of fire and war. More recently the older portions of these county records have been centralized in state archives. A genealogical guide to land records at the county, state (or colony), and national levels is presented in The Land Records of America and their Genealogical Value, by E.K. Kirkham (by the author, 1963).

The records pertaining to land survey and disposal in the public land states are deposited in various offices of the National Archives (Center for Cartographic and Architectural Archives, Legislative and Natural Resources Branch, General

Archives Division, and regional archives branches) and various offices of the Bureau of Land Management (Eastern States Office, Denver Service Center, and individual state land offices). Unfortunately, there is no comprehensive guide to these records. A summary overview of the land survey records (instructions, plats, and field notes) in the National Archives, state archives, and the Eastern States Office was presented by Bureau of Land Management employee L.J. Bouman in "The Survey Records of the General Land Office and Where They Can Be Found Today," American Congress on Surveying and Mapping Proceedings of 36th Annual Meeting (February 1976), pp. 263-72. The availability of township plats and field notes at the county, state, and Federal levels, is also discussed by geographer W.D. Pattison in "Use of the U.S. Public Land Survey Plats and Notes as Descriptive Sources," Professional Geographer 8 (January 1956): 10-14. He also demonstrates the utility of these records for historical geographical research by citing numerous secondary articles and dissertations produced from the 1930s to 1950s. In discussing the records created by the Bureau of Land Management, archivist J. Smith divides the records into four categories (survey, status, case, and legal records) in "Settlement of the Public Domain as Reflected in Federal Records: Suggested Research Approaches," pp. 290-304 in Pattern and Process, edited by R.E. Ehrenberg (Howard University Press, 1975). Another archivist, R.S. Maxwell discusses only the land disposal records (land entry papers, tract books, monthly abstracts, patents and township plats) in Public Land Records of the Federal Government, 1800-1950, and their Statistical Significance, Reference Information Paper No. 57 (National Archives and Records Service, 1973). The land entry papers, which are the most voluminous of these land disposal records, are listed in Preliminary Inventory of the Land Entry Papers of the General Land Office, Preliminary Inventory No.22, by H.P. Yoshpe and P.P. Brower (National Archives, 1949). The research potential of the land entry papers is discussed by R.W. Harrison in "Public Land Records of the Federal Government," Mississippi Valley Historical Review 41 (September 1954): 277-88, and R.S. Lackey in "The Genealogist's First Look at Federal Land Records," Prologue 9 (Spring 1977): 43-45. Land records in the National Archives' San Francisco Federal Archives and Records Center are discussed by J.P. Heard, "Resource for Historians: Records of the Bureau of Land Management in California and Nevada," Forest History 12 (July 1968): 20-26. Early land records, particularly for Ohio, Indiana, Illinois, Alabama, Mississippi and Louisiana are abstracted in the first three volumes of Federal Land Series, compiled by C.N. Smith (American Library Association, 1972-1980).

Federal land records have been used by historical geographers for a number of research topics. Those using the survey records have emphasized the development of the rectangular survey system or the effects of the survey system on the landscape. An example of the former is Beginnings of the American Rectangular Land Survey System, 1784-1800, Department of Geography Research Paper No. 50, by W.D. Pattison (University of Chicago, Department of Geography, 1957). The second type is exemplified by Original Survey and Land Subdivision: A Comparative Study of the Form and Effect of Contrasting Cadastral Surveys, Association of American Geographers Monograph Series No.4, by N.J.W. Thrower (Rand McNally and Company, 1966), which shows the contrast between the rectangular and metes and bounds survey systems on the Ohio landscape. Order Upon the Land: The U.S. Rectangular Land Survey and the Upper Mississippi Country, by H.B. Johnson (Oxford University Press, 1976), demonstrates the difficulty of imposing the rectangular survey system on the hilly topography of southeastern Minnesota and northwestern Wisconsin. A good example of a geographer's use of the land disposal records is C.B. McIntosh's study of the Nebraska Sand Hills, which is reported in three articles in the Annals of the Association of American Geographers: "Patterns from Land Alienation Maps," 66 (December 1976) :570-82; "Use and Abuse of the Timber Culture Act," 65 (September 1975): 347-62; and "Forest Lieu Selections in the Sand Hills of Nebraska," 64 (March 1974): 87-99. Both the land survey and land disposal records are used in The Initial Evaluation and Utilization of the Illinois Prairies, 1815-1840, Department of Geography Research Paper No. 64, by D.R. McManis (University of Chicago, Department of Geography, 1964) to reconstruct vegetation patterns and to determine settlers' initial perception of woodland and prairies in choosing settlement sites.

Although states such as Florida, Louisiana, New Mexico, Arizona, Colorado, and California are public land states, the original land holdings in these states originated under Spanish and Mexican governments. When the United States government acquired these territories, the General Land Office agreed to honour title to lands granted by the Spanish and Mexican governments. Numerous records were accumulated in the process of proving title to these private land claims. For example, the records associated with the California private land claims, which are in the California State Land Office, Bancroft Library, and the National Archives, are indexed in Index of the Spanish-Mexican Private Land Grant Records and Cases of California, by J.N. Bowman (University of California, Bancroft Library, typescript, 1958). The role of these land grants in the settlement of California is examined by geographer D. Hornbeck in "Land Tenure and Rancho Expansion in Alta California, 1784-1846," Journal of Historical Geography 4 (October 1978): 371-90, and "The Patenting of California's Private Land Claims, 1851-1885," Geographical Review 69 (October 1979): 434-48. In the former article he discusses the various types of records that relate to the California private land claims. Microfilmed copies of records pertaining to New Mexico's private land claims are described in A Guide to the Microfilm of Papers Relating to New Mexico Land Grants, by A.J. Diaz (University of New Mexico Press, 1960).

Land records for the nonpublic land states were normally created and maintained in the state. In the case of the original thirteen states, most of the land was disposed of during the colonial period, the records were maintained by a variety of colonial administrators, and many of the records, especially survey plats, are no longer extant. Virginia's land records are described in Virginia Land Office Inventory, compiled by D. Gentry (Virginia State Library, 1975). These records, which are in the custody of the Virginia State Library, consist primarily of patents, but also include survey plats and abstracts. The records date from the colonial and commonwealth periods and include the records of the Northern Neck Proprietary. Seventeenth-century land grants are abstracted in Cavaliers and Pioneers: Abstracts of Virginia Land Patents and Grants, 1623-1666, edited by N.M. Nugent (Dietz Press, 1934; reprinted, Genealogical Publishing Co., 1969). In Georgia the land records are in the custody of the Surveyor-General Department. This office has issued nine booklets, English Crown Grants in Georgia, 1755-1775, compiled by P. Bryant and M.B. Hemperley (1972), which abstract the royal land grants for eleven Georgia parishes.

Besides the land records that document the initial survey and sale of land at the national and state level, there are the county records, such as deed books, which document the subsequent transfer of land. Obviously, there is no comprehensive guide to these records, although some states have centralized these records in a state land office or archival agency. Most county records are available on microfilm from the Genealogical Society Library, Salt Lake City. Closely associated with the county land records are the probate records which are also maintained at the county level. The probate records, which usually include wills, inventories, and accounts of administration, list the extent of a person's real estate, personal property, and other assets at the time of death. The use of probate records in colonial historical research has been discussed by G.L. Main in two articles: "Probate Records as a Source for Early American History," William and Mary Quarterly, 3rd series, 32 (January 1975): 89-99, and "The Correction of Biases in Colonial American Probate Records," Historical Methods Newsletter 8 (December 1974): 10-28. Another useful article is D.S. Smith, "Underregistration and Bias in Probate Records: An Analysis of Data from Eighteenth-Century Hingham, Massachusetts," William and Mary Quarterly, 3rd series, 32 (January 1975): 100-110. Several historical geographers working in the colonial period have made extensive use of county land and probate records to reconstruct landownership, wealth and social status, household composition, and agricultural activity. This research includes The Evolution of a Tidewater Settlement System: All Hallow's Parish, Maryland, 1650-1783, Department of Geography Research Paper No. 170, by C.V. Earle (University of Chicago, Department of Geography, 1975); The Best Poor Man's Country: A Geographical Study of Early Southeastern Pennsylvania, by J.T. Lemon (Johns Hopkins University Press, 1972); and Commercialism and Frontier: Perspectives on the Early Shenandoah Valley, by R.D. Mitchell (University Press of Virginia, 1977).

Using county landownership records to reconstruct landownership patterns for any interval subsequent to the initial sale can be a tedious research process. Rent rolls or assessment lists, when available, can aid in this type of research. Another type of record that provides a similar shorthand approach to reconstructing landownership patterns is the commercially published county landownership maps and plat books, which were quite prevalent during the last half of the nineteenth century and the early twentieth century. The collection of county landownership maps in the Library of Congress is listed in Land Ownership Maps: A Checklist of Nineteenth Century United States County Maps in the Library of Congress, compiled by R.W. Stephenson (Library of Congress, 1967), while their collection of county atlases or plat books is included in United States Atlases: A List of National, State, County, City, and Regional Atlases in the Library of Congress, compiled by C.E. LeGear (Library of Congress, 1950; reprinted, Arno Press, 1971). An excellent example of a union listing of county atlases for one state is Windows to the Past, A Bibliography of Minnesota County Atlases, by M. Treude (University of Minnesota, Center for Urban and Regional Affairs, 1980). N.J.W. Thrower discusses the research potential of the county atlases in "The County Atlas of the United States," Surveying and Mapping 21 (September 1961): 365-73, while a test case of their accuracy is presented by O.F.G. Sitwell in "County Maps of the Nineteenth Century as Historical Documents: A New Use," Canadian Cartographer 7 (June 1970): 27-41.

## Travel and Exploration Accounts

A traditional source for information in historical geography is the descriptive account found in travellers' diaries and explorers' reports. Geographers have traditionally used this data as evidence for reconstructing physical landscapes, settlement patterns, and economic activity. More recently, they have also dealt with environmental perception as portrayed in these accounts.

Travellers' diaries, which usually record a traveller's first or only impression of an area, are available in manuscript form in a wide range of archival and historical manuscript collections. Many are also available in a published format so that a larger audience may have access to them. Over 5,000 manuscript diaries in 350 depositories are described in American Diaries in Manuscript, 1580-1954: A Descriptive Bibliography, by W. Matthews (University of Georgia Press, 1974), while published diaries are listed in American Diaries: An Annotated Bibliography of American Diaries Written Prior to the Year 1861, by W. Matthews (University of California Press, 1945). There are also several bibliographies that list diaries and travel accounts by broad regions or individual states. Over 2,700 diaries pertaining to the southeastern United States are included in a six volume set published by the Oklahoma University Press. This set was published under three titles: Travels in the Old South: A Bibliography, edited by T.D. Clark (3 vols., 1956-59); Travels in the Confederate States: A Bibliography, edited by E.M. Coulter (1948); and Travels in the New South: A Bibliography, edited by T.D. Clark (2 vols., 1962). Travel literature pertaining to the midwestern states is listed in Early Midwestern Travel Narratives: An Annotated Bibliography, 1634-1850, by R.R. Hubach (Wayne State University Press, 1961). Examples of state bibliographies include Travelers in Tidewater Virginia, 1700-1800: A Bibliography, by J. Carson (University Press of Virginia for Colonial Williamsburg, 1965); New Jersey in Travelers' Accounts, 1524-1971: A Descriptive Bibliography, by O.S. Coad (Scarecrow Press, 1972); and Travelers in Texas, 1761-1860, by M.M. Sibley (University of Texas Press, 1967). Sources of travel literature, including Yale University, Newberry and Bancroft Libraries, are discussed in nine of the thirteen essays in Travelers on the Western Frontier, edited by J.F. McDermott (University of Illinois Press, 1970). These essays were originally presented at the Conference on Travelers on the Western Frontier, which was held at the University of Southern Illinois, Edwardsville, in February 1968.

The validity and utility of travel narratives for historical geographical researcharediscussed in the several articles by H.R. Merrens that were cited earlier. Another useful analysis of travel accounts, in terms of their perception of the landscape, is presented by R.C. Bredeson in "Landscape Description in Nineteenth-Century American Travel Literature," American Quarterly 20 (Spring 1968): 86-94. The deception found in some supposedly authentic travel accounts is the main theme in Travelers and Travel Liars, 1660-1800, by P.G. Adams (University of California Press, 1962).

A significant subset of travel literature includes the reports prepared by government expeditions during the nineteenth century. Although these sources have their own inherent biases (military reconnaissance and economic exploitation), they have the potential of presenting a more factual representation than the private accounts which may suffer from exaggeration, untruth, or literary license. Exploration accounts that were published in the Congressional Serial Set are listed in Reports of Explorations Printed in the Documents of the United States Government: A Contribution Toward a Bibliography, by A.R. Hasse (Government Printing Office, 1899). Another useful bibliography, Catalogue and Index of the Publications of the Hayden, King, Powell, and Wheeler Surveys, by L.F. Schmeckebier (Government Printing Office, 1904; reprinted, Da Capo Press, 1971) lists the published annual reports, final reports, bulletins, miscellaneous publications, and maps that resulted from the four government surveys of the western United States, 1867-1879. In a more recent article, "Legacy of the Topographical Engineers: Textual and Cartographic Records of Western Exploration, 1819-1860," Government Publications Review 7A (1980): 111-16, F.N. Schubert, an historian with the Office of the Chief of Engineers, describes the major bibliographical guides to the textual and cartographic records produced by the Army Topographical Engineers. The Engineers' records, as well as the cartographic and textual records of other major Federal mapping and surveying agencies are also described in Geographical Exploration and Mapping in the Nineteenth Century: A Survey of the Records in the National Archives, Reference Information No.66, by R.E. Ehrenberg (National Archives and Records Service, 1973). The utility of these government exploration accounts for historical geographical research has been addressed by H.R. Friis in several essays: "The Documents and Reports of the United States Congress: A Primary Source of Information on Travel in the West, 1783-1861," pp. 112-67, in Travelers on the Western Frontier, edited by J.F. McDermott (University of Illinois Press, 1970); "The Image of the American West at Mid-Century (1840-60): A Product of Scientific Geographical Exploration by the United States Government," pp. 49-64, in The Frontier Re-examined, edited by J.F. McDermott (University of Illinois Press, 1967); and "Original and Published Sources in Research in Historical Geography: A Comparison," pp. 139-59, in Pattern and Process, edited by R.E. Ehrenberg (Howard University Press, 1975). The history of the records associated with one of the earliest expeditions sponsored by the Federal government is reviewed in A History of the Lewis and Clark Journals, by P.R. Cutright (University of Oklahoma Press, 1976).

Geographers have found numerous uses for travel and exploration accounts. J.A. Jakle reconstructed various images of the Ohio River Valley based on approximately 400 travel accounts in Images of the Ohio Valley: A Historical Geography of Travel, 1740 to 1860 (Oxford University Press, 1977). He also evaluated the biases inherent in travel accounts with particular emphasis on different modes of transportation. J.L. Allen utilized exploration accounts, particularly those associated with the Lewis and Clark expedition, to analyse the relationship between the exploratory process and the creation of geographic images of the American West in Passage Through the Garden: Lewis and Clark and the Image of the American Northwest (University of Illinois Press, 1975). Travellers' diaries and explorers' accounts were some of the sources used by M.J. Bowden in reconstructing the popular image of the Great Plains in "The Great American Desert and the American Frontier, 1800-1882: Popular Images of the Plains," pp. 48-79 in Anonymous Americans: Explorations in Nineteenth-Century Social History, edited by T.K. Hareven (Prentice-Hall, 1971). Similarly, M.P. Lawson used climatic observations recorded in travellers' diaries as one source in reconstructing the 1849 climatic conditions in The Climate of the Great American Desert: Reconstruction of the Climate of Western Interior United States, 1800-1850 (University of Nebraska Press, 1974).

Commercial Records

The final grouping of records is a miscellaneous grouping of materials that are termed commercial and industrial records. Their similarity is that these records document a business or industrial operation or were the primary product of a commercial venture: newspapers, business and city directories, and fire insurance maps.

A major source of nineteenth-century industrial statistics is the U.S. Census Bureau's census of manufactures. The development of the manufacturing censuses and the content of the respective reports are reviewed in The Censuses of Manufactures, 1810-1890, Reference Information Paper No.50, by M.H. Fishbein (National Archives and Records Service, 1973). The value and accuracy of the mid-nineteenth century manufacturing returns are discussed in two articles by M. Walsh in Historical Methods Newsletter: "The Census as an Accurate Source of Information: The Value of Mid-Nineteenth Century Manufacturing Returns," 3, no.4 (September 1970): 3-13, and "The Value of Mid-Nineteenth Century Manufacturing Returns: The Printed Census and Manuscript Census Compilations Compared," 4 (March 1971): 43-51. Her study of industrial development in Wisconsin, The Manufacturing Frontier: Pioneer Industry in Antebellum Wisconsin, 1830-1860 (State Historical Society of Wisconsin, 1972), and R.G. LeBlanc's Location of Manufacturing in New England in the 19th Century (Dartmouth College, Geography Department, 1969) are good examples of the use of this source.

Archives and manuscript collections relating to individual industries and businesses are another source for reconstructing past spatial patterns of economic activity. The archival holdings of over 200 private businesses are described in Directory of Business Archives in the United States and Canada (Society of American Archivists, 1980). One repository that is particularly noted for its holdings of business and industrial archives is the Eleutherian Mills Historical Library near Wilmington, Delaware. A guide to its holdings has been published as A Guide to the Manuscripts in the Eleutherian Mills Historical Library, Accessions through the Year 1965, by J.B. Riggs (Eleutherian Mills Historical Library, 1970). Accessions for the years 1966 to 1975 are listed in a supplement published in 1978. Records pertaining to colonial commercial firms are scattered in numerous repositories, but a number of merchants' accounts and business papers are found in the Research Library, Colonial Williamsburg. These records are listed in Guide to the Manuscript Collections of Colonial Williamsburg, compiled by M.G. McGregor (2nd edn., Colonial Williamsburg, 1969). Several articles describing the research potential of individual business archives have been written by geographers. Based on his research in the archives of the Union Pacific and Northern Pacific Railroad Companies, D.W. Meinig describes the types of records that are available for the study of the historical geography of the railroads in "Railroad Archives and the Historical Geographer," Professional Geographer 7 (May 1955): 7-10. The Hudson's Bay Company's archives have been used by several geographers studying the fur trade in Canada and the northern Great Plains. Assessments of these records for geographical research are presented by D.W. Moodie in "The Hudson's Bay Company's Archives: A Resource for Historical Geography," Canadian Geographer 21 (Fall 1977): 268-74, and A.J. Ray in "The Early Hudson's Bay Company Account Books as Sources for Historical Research: An Analysis and Assessment," Archivaria 1, no.1 (Winter 1975/6): 3-38, and "The Hudson's Bay Company Account Books as Sources for Comparative Economic Analysis of the Fur Trade: An Examination of Exchange Rate Data,"Western Canadian Journal of Anthropology 6 (1976):44-50.

In addition to business records that provide information about the spatial aspects of a particular industry, there are certain commercial enterprises that result in an end product that has secondary research value. The newspaper industry is a good example because the primary function of the newspaper is the dissemination of news and editorial opinion. While these elements of the newspaper have been used widely to study the development of political and public opinion, other aspects of the newspaper publishing industry, such as advertisements, specialized newspapers, or circulation statistics, provide useful information about commerce, manufacturing, and agriculture. Several articles discuss the varied research potential of newspapers. For example, P.O. Wacker evaluates newspaper advertisements from colonial New Jersey as a source of economic and cultural data in "Historical Geographers, Newspaper Advertisements, and the Bicentennial Celebration," Professional Geographer 26 (February 1974): 12-18. The content of agricultural newspapers is discussed by R.T. Farrell, "Advice to Farmers: The Content of Agricultural Newspapers, 1860-1910," Agricultural History 51 (January 1977): 209-17, while the use of newspapers in immigration studies is evaluated by R. Hayward and B.S. Osborne in "The British Colonist and the Immigration to Toronto of 1847: A Content Analysis Approach to Newspaper Research in Historical Geography", Canadian Geographer 17 (Winter 1973): 391-402. The availability and limitations of newspaper circulation statistics for central place studies are demonstrated by R.E. Preston in "Audit Bureau of Circulations Daily Newspaper Records as a Source in Studies of Post-1915 Settlement Patterns in the United States and Canada," Historical Geography Newsletter 7 (1977): 1-12.

A wide range of newspapers is available in research libraries and on microfilm. A bibliography of pre-1820 newspapers is found in History and Bibliography of American Newspapers, 1690-1820, by C.S. Brigham (American Antiquarian Society, 1947; reprinted 1962), while post-1821 papers are listed in American Newspapers, 1821-1936: A Union List of Files Available in the United States and Canada, by W. Gregory (H.W. Wilson Co., 1937). Newspapers that are available on microfilm are listed in Newspapers in Microform: United States, 1948-1972 (Library of Congress, 1973). Subsequent additions to this listing were published in Newspapers in Microform: United States, 1973-1977 (1978).

City and business directories, which provide a listing of residents and/or businesses in a particular city or geographical area, provide a useful complement to the manuscript census schedules in studies of population mobility and social or economic characteristics. Over 1,600 directories, most of which are available in the American Antiquarian Society Library, are listed in Bibliography of American Directories Through 1860, by D.N. Spear (American Antiquarian Society, 1961; reprinted, Greenwood Press, 1978). City directories for the 300 largest U.S. cities are listed in A Handy Guide to Record-Searching in the Larger Cities of the United States, by E.K. Kirkham (Everton Publishers, Inc., 1974). The content, biases, and research potential of city directories are reviewed by P.R. Knights in "City Directories as Aids to Ante-Bellum Urban Studies: A Research Note," Historical Methods Newsletter 2, no.4 (September 1969): 1-10. M.P. Conzen describes the research potential of state business directories in "State Business Directories in Historical and Geographical Research," Historical Geography Newsletter 2, no.2. (Fall 1972): 1-14.

A third research tool, fire insurance maps, are prepared primarily for insurance underwriters in determining the value risk of insured properties. Historical researchers have found these maps to be a valuable aid in historical urban research, especially when used in conjunction with city directories and manuscript census schedules. The development and intended use of fire insurance maps is described by W.W. Ristow in "United States Fire Insurance and Underwriters Maps, 1852-1968," Quarterly Journal of the Library of Congress 25 (July 1968): 194-218. A union listing of fire insurance maps produced by the Sanborn Company, the major publisher of fire insurance maps, is presented in Union List of Sanborn Fire Insurance Maps Held by Institutions in the United States and Canada, by R.P. Hoehn et al.(2 vols., Western Association of

Map Libraries, 1976-77). Two large collections of fire insurance maps are in the Library of Congress and the Geography Department, California State University, Northridge. These collections are listed in Fire Insurance Maps in the Library of Congress: Plans of North American Cities and Towns Produced by the Sanborn Map Company (Library of Congress, 1981), and A Catalogue of Sanborn Atlases at California State University, Northridge, by G. Rees and M. Hoeber (Western Association of Map Libraries, 1973). The use of fire insurance maps in historical geographical research is discussed in the following articles: S.H. Ross, "The Central Business District of Mexico City as Indicated on the Sanborn Maps of 1906," Professional Geographer 23 (January 1971): 31-39; L.J. Gibson, "Tucson's Evolving Commercial Base, 1883-1914: A Map Analysis," Historical Geography Newsletter 5, no.2. (Fall 1975): 10-17; and E.N. Moody, "Urban History in Fire Insurance Maps: Nevada as a Case Study," Western Association of Map Libraries Information Bulletin 10 (March 1979): 129-38.

Although population, property, travel, and commercial records have been employed successfully in research in historical geography, the selection of these records for review does not mean to imply that they are the only records available to geographers. No historical or archival source can be claimed exclusively by one discipline, nor should there be any limit to the materials that are of potential use to geographers. Consequently, researchers using these sources will need to be aware of the literature that describes individual institutional holdings and that which describes and critiques various types of records.